OTHER TITLES BY JOHN JANOVY JR.
AVAILABLE IN BISON BOOKS EDITIONS

Dunwoody
Reflections on the High Plains
Pond
Wetlands and the Cultivation

of Naturalists

John Janovy, Jr.

With a new epilogue by the author

UNIVERSITY OF NEBRASKA PRESS
LINCOLN AND LONDON

First Bison Books printing: 2001

Library of Congress Cataloging-in-Publication Data
Janovy, John, 1937–
Dunwoody Pond: reflections on the high plains wetlands and the
cultivation of naturalists / John Janovy, Jr.
p. cm.
Originally published: New York: St. Martin's Press, c1994. With new
epilogue.
Includes bibliographical references.
ISBN 0-8032-7616-8 (pbk.: alk. paper)
1. Natural history—Nebraska. 2. Wetlands—Nebraska. 3. Ponds—
Nebraska. 4. Pond ecology—Nebraska. 5. Naturalists—Nebraska.
I. Title.
QH105.N2 J344 2001
508.782'89—dc21
00-066595

*To Tammy, Rich, Marv, Aris, Laura,
and all the others who have gone before*

Contents

Foreword

This book is about parasites, the scientists who decide to study them, and the places where these scientists work. The parasites are those that live in or on beetles, damselflies, frogs, toads, fish, near-microscopic crustaceans, and similar kinds of animals. To most people, these parasites are not very important ones; that is, they don't cause human diseases. They are, however, extraordinarily interesting beasts because they live convoluted lives and have intricate structures. The people who decide to study such creatures also live interesting and convoluted lives, especially in their minds. These humans constitute a privileged, but impoverished, class. Nobody can ever take away from them the heady rush that comes from original discovery.

Like all naturalists, they have strong emotional ties to the field sites where they do their research. Long after they've walked away from the landscapes described herein, they will remember the nondescript ponds, streams, and marshes with an increasingly nostalgic sense that those years of intellectual struggle were among the best of their years on Earth. So this book is also about the link between human beings and the physical locations where they mold themselves into practicing scientists. There is a lesson to be learned from watching this process: That link can never be broken, so choose carefully, if you can, the place where you decide to become well educated.

This book is also about death, in the sense that one usually must dissect animals in order to study the other animals living inside them. All of the young scientists in this book have struggled mightily with this aspect of their work. As a consequence, their chosen problems are always a compromise between curiosity, the need for money and time, their self-imposed requirements for a large amount of observations, easily accessible animals, conceptually important research projects, and the public's emotional reaction to their science. From my own student days, I don't remember being castigated for killing animals in order to learn something of their nature. I'm sure there were people who felt that what I was doing was "wrong," but they didn't break into my laboratory and set my mosquitoes free. But the negative emotional reaction that nowadays often surrounds a scientist's work is compounded when the scientist kills "pretty" animals in order to study other animals that people generally hate, i.e., parasites. This negative emotional reaction is baggage that my students try to discard early in their careers, usually by choosing host animals (those that contain their beloved parasites) that are small, ugly, nonfurry, and exceedingly plentiful. If the chosen species can be raised in large numbers in the lab, then so much the better. That is, human negative emotional reaction toward one's research is countered most easily by two means: doing research on problems of immediate economic importance, or, by doing research on animals that look and act the least like humans.

The most human-like animals dissected by the people in this book are toads. I love toads, but I don't get very upset by a student's doing a legitimate scientific research project that requires the sacrifice of *many* toads of a common species. When I first decided to become a professional scientist, I approached my ornithology teacher to ask whether he would guide my research. He didn't answer yes or no, but launched into a discourse on the collecting of things you love in order to study them seriously. The "things" of course, would have been birds. The collections would have required the use of guns and nets. Now I find myself in his position, rationalizing the use of animals to study other animals. In the case of my beloved toads, it still bothers me to kill one, but it bothers me more to have to answer to those who say it should not be done in the name of science, especially when that science involves the study of parasites. One garter snake can eat more baby toads than all my graduate students collectively have used in their research in the last decade. I won't tally the cost in baby toads, of prairie river diversion for the purposes of irrigation; the numbers are astronomical. Long ago I came to the conclusion that it was all right to study parasites that live in toads, and that my honest scientific use of toads in order to feed one student's mind or raise a new crop of scientists, was as worthy an endeavor as the feeding of one garter snake or the raising of one crop of irrigated subsidized corn.

This book is also about all-consuming problems. The hallmark of science is its ability to compete with the more mundane aspects of life, as well as with arts and letters, for your thoughts and dreams. Such competition is especially strong when you're surrounded by aggressive articulate people working on problems you personally find fascinating. My laboratory has been filled with such students. We spent our days hunkered down over a microscope or a computer terminal, but used our nights to watch the storms from a place called Monkey Rock and talk about everything from cosmology to beetle guts, from the teaching of biology to war, politics and auto mechanics. The enormity of this privilege—that of being fortunate enough to associate with such people and their ideas—must have sunk in finally and awakened me, literally as well as metaphorically. This book, therefore, is not only about parasites and death, it's also a thank you note to my students for the use of their time and talents, as well as a reminder to myself of the many things we talked about—things that need to be written down before they are lost from memory.

Students are what makes academic science one of the truly noble of all professions. I've always been surprised at the way some of my colleagues shun graduate students, or struggle to recruit them, or ultimately grind them down. Potential students have always walked in my door, looked around the room, then asked if there were an opportunity to do some serious study, about almost anything. I have no idea why they do it, but over the years, there has been a steady stream of them, all unique in unique ways, all highly unusual people, some ultimately more successful than others, but nevertheless very capable and intelligent human beings bent on some mission they weren't always willing to reveal—completely. Not long ago I mentioned this phenomenon to my own graduate adviser, Dr. J. Teague Self, long retired from the faculty at the University of Oklahoma. "John," he said, "they always accused me of stealing the best graduate students." Then he laughed and said, "But all I did was go into the lab and work, and they came in." Yes, Dr. Self, I thought, I did exactly the same thing—just walked into your office one day and asked if you would give me some guidance in the study of animals. Maybe if we knew why people sought such help, then we'd have a clue about the origin of scientists.

The individuals who inhabit the following pages are ones that in general have simply walked into my life with the intent of becoming parasitologists—people who study animals that live inside other animals, e.g., the tapeworms in your family pet dog. Their stories are important ones to tell for a wide variety of reasons. First, the United States of America desperately needs the human resources these students represent. As a nation, and a culture, we are not in very good shape, intellectually speaking. We are a people with vast natural resources and a vast appetite.

We command a lion's share of the global energy supplies. We've built a massive technological infrastructure, and handed humanity's most powerful weapons over to the politicians. Our science has served us very well, if one considers only the developments that help us exploit and alter our environment. But technology now keeps us alive for so long, and in such frail condition, that our philosophers argue the merits of euthanasia, living wills, and assisted suicide. Technology lets us handle human sperm and eggs in ways that completely redefine procreation, in the process spawning a legion of attorneys who adjourn to a local watering hole to discuss whether a fertilized ovum, nestled in a vial in a tank of liquid nitrogen, should be considered property, or a child, within the context of a divorce settlement. Technology shields us from the cold, lifts us into the sky for a trip to grandmother's, floods our living rooms with the intimate details of life and death half a world away, opens a window on far galaxies, and turns fantasy into reality at a hellish pace. Technology competes with art for the role of civilization's icon.

Yet we hang on to a certain mysticism, a certain fatalism, born, sometimes I think, out of an unquestioning acceptance of Western Christianity with its absolute rightness and wrongness, absolute causality, presupposition of sinfulness, and apocalyptic prophecies. It is for this reason that I feel we are in such need of the people you'll meet in the following pages. The ends of easily identifiable blocks of time have always been accompanied by social stress. December holidays are notorious for their depressing effects. The end of the First Millennium saw great turbulence stirred at least in part by Biblical predictions. Once again we are within a few short years of the end of a millennium. Voices scream from my newspaper and television set, generously distributing blame, confidently proclaiming disaster for all humanity. But the people who've walked into my laboratory are rational beings who have enormous faith in their own very human talents, and little use for prophets. They attack gigantic problems with only their minds and hands. They are models for a type of human being that takes pride in its brain, in its ideas, instead of in its weapons or power. So I've set about to tell their stories. We need to know where these kinds of people come from, how they are shaped, and how they think, with the hope that in the telling, we'll discover how to generate some more of them.

Their stories also need to be told because they represent the finest products of the American formal educational enterprise. Our schools have been our battleground, both figuratively and literally, for a generation. If media reports have any truth to them whatsoever, there is little hope this situation will change. We daily read of stabbings, shootings, beatings, narcotics, failures, racial stress, teacher strikes, parental rebellions, all painted with the haze of a grand moral debate over prayer, television, art, rock lyrics, sexual orientation, abortion, and family planning. A sociologist would look to one tiny brief event in our history as a summary of the plight of our public schools: A newly elected President decides to send his teenage child to a private school and the headlines scream that he's violated a sacred symbolic promise. What the hell difference does it make where a President of the United States and his wife decide to send their child to school? None. Absolutely none. Routinely the public schools produce people who walk into my classes, and into my laboratory, seeking intellectual challenges and opportunities to explore in depth a tiny segment of the natural world. These people scan the daily headlines and ask me: What the hell difference does it make where a President's kid goes to school? I don't have to answer. The answer is already implied in the tone of their voices: none, absolutely none. But that same tone always contains a plea: Why can't we grow up, become mature, intellectually, as a nation?

I cannot answer the second question. I do not know how to cure all the ills of America's educational system, and this book is not an attempt to do that. But I do know how to tell the stories of ordinary young people who've succeeded admirably by coming up through that system. These particular individuals are probably more libertarian than they should be, but they also read the editorial pages, listen to National Public Radio, watch C-Span and CNN, read highly technical literature, sophisticated nonfiction books and complicated novels, stand up in front of national scientific meetings and present the results of their original investigations, spend many of their nights and weekends on government business, discuss ideas rather than gossip about the private lives of their colleagues, can generally distinguish fact from fairy tale, publish serious papers in anonymously reviewed journals, acquire needed computer skills by reading and practicing, apply statistical methods in order to uncover their own false perceptions, and serve as role models for the hundreds of others they teach. Their average age is about twenty-four.

In other words, the main characters in this book are all young scientists. The job market for their skills is not very promising but that situation does not deter them very much. Although they're still young, they've seen much intellectual ability wasted by people choosing a paycheck over a scientific puzzle, and they are trying desperately to avoid that trap. For this reason, I think, we need to ask: Where do young biologists come from, in general, as opposed to individually and specifically? In the years since I last wrote about Keith County an enormous amount of raw talent has passed through my office and laboratory. I saw the beginnings of careers, rather than the endings, the starts of these journeys rather than the biographies of Nobel Prize winners, the first commitments to a life that centers around difficult questions, instead of the last look at a paper submitted, accepted, and published long ago. My goal in writing this book is to humanize science, bring it home to the kid next door, make it the property of the masses. But by "science" I don't mean formalized science, applied science, and technology. Instead, I define "science" as a state of mind, a "way of knowing," to quote the title of John

Moore's magnificent book. If there is a personal crusade evident in these pages, it's my dream of being able to use the word "evolution" in a lecture to three hundred students in an introductory biology class without having someone walk out. The people in the following pages are the ones who stayed to listen, then went to the lab, and to the museum, and to the field, to gather their own observations of the world in which they live.

The setting for their stories is western Nebraska, at a place we often call "Camelot" but which is officially known as the Cedar Point Biological Station. All of the sites described herein are within a few miles of CPBS. The fact that this mini-campus is on the edge of the High Plains is not of particular importance, however; this book could easily have been written about any place where young people go, on purpose, knowing exactly what they're doing, in order to mold themselves into serious scholars. My guess is that the large cities are filled with equivalent habitats that produce a wide diversity of highly educated individuals. My hope is that the stories that follow will encourage my fellow teachers across the nation to rekindle the flame of higher education as a driving force in American culture, by tying that education, emotionally, to the landscape in which it occurs.

—JOHN JANOVY, JR.

Part I

Students at Work

*A*ris, Tami, Laura, and Marv are not completely at ease with what I've written, although all have read their stories, made a few comments, and dutifully signed their releases. At the time they were working so many hours a day, getting so dirty, struggling with their decisions, they didn't think of themselves as heroes, role models, or historical figures. But now they've been cast in words, fixed in time. People they've never met may write them letters. Years hence, they'll get introduced to someone who knew them when they were young, thinks of them always as young, at the beginnings of their intellectual journeys. These strangers will have read the chapters and will want to know how to find Dunwoody Pond, or Nevens Ranch. I don't know what Aris, Tami, Laura, and Marv will be doing when that happens, or how they will answer the request for directions. My guess is Aris will say: "Dunwoody Ponds are everywhere; you just have to find one of your own." Tami will tell the stranger to ask permission, drive slowly through the yard, and pay for the visit with some interesting conversation. Laura will answer: "It's way down a long, long road." Marv will get out the map and give a set of detailed instructions on how to get there; then he'll ask that you go somewhere else.

1
Dunwoody Pond

Sand Rivers

— PETER MATTHIESSEN

Winds came out of the northeast, spawned from the glacier's face. They blew for weeks, bearing dirt, sometimes pushing dark clouds across the sun and moon, and whistling, always whistling, through the nights. The weeks passed into years and still the winds blew. Sometimes the noon sky grew black with ashes from a far off volcano. Strange animals—striped horses, antelopes with two sets of horns, elephants with four tusks—turned their eyes away from the stinging grit and waded into shallow potholes; herons stepped aside and let the animals drink. Sand fell from the sky, gathered like snow in drifts, and then was thrown into waves that marched across the land. As the centuries passed, the dunes became covered with plants that thrived in dry landscapes, blistering heat, and searing cold. But all the strange animals eventually died. Today, instead of mastodonts, it's Duane Dunwoody who stands in the July sun and surveys the surrounding prairies.

"You need to meet my mother-in-law, Doc." He's drinking instant coffee, eating a bread and butter sandwich. He offers me a drink of the former and a bite of the latter; I accept the bite. "Mom has this collection of pressed flowers. Goes way back. She's got plants that not even the experts know lived around here."

That doesn't surprise me very much. Experts nowadays are more likely to be searching for money in Washington than prairie flowers in Dunwoody's pasture.

"Sure you don't want a drink o' my coffee? I don't have AIDS."

Two miles down a sand and gravel road, north of a town with no grocery store, in the middle of the sixty-eighth most populous county of the thirty-fourth most populous state, Duane Dunwoody is completely aware of the social and political issues of our day and the way these issues permeate the next generation's thoughts. His wisecrack about Acquired Immune Deficiency Syndrome is testimony to the strength of that electronic net now binding five billion of us into a single, tight, information-sharing community, regardless of our personal desires, and needs, for elbow room.

"I've had enough coffee this morning for ten people, Mr. Dunwoody." I let the AIDS remark drop. "Yes, I'd like to meet your mother-in-law. I'll come back some time."

"You asked about another pond, Doc." He hitches up his pants. "I told you about that one over on Buckhorn Springs, but my wife reminded me there's one right over here on th' other side of the tracks. It's been there twenty years."

His sweeping gesture includes not only the tracks, but also several houses, a barn, a low wooden hangar, and a moving van, a semi-trailer, with orange paint peeling. Among the houses swept up in the arc of Dunwoody's arm is a beautiful, yellow, recently refurbished Victorian two story—his mother's former residence, now the home of his daughter. The daughter and her family moved around so much, the story goes, that Dunwoody finally told them just to buy the moving van and keep their furniture in it. That way they wouldn't have to do so much packing, storing, and unpacking. Some would say the giant trailer is a monument to eccentricity. Others would claim it's a symbol of practicality. Still others might argue that in today's world, Dunwoody's level of practicality is what's eccentric. But eccentricity is no barrier to success; he owns his land free and clear, and wakes up every morning to songs of the grasslands instead of gunfire.

"Yes, I'd like to find another pond just like this one," I say, out of courtesy; there are no other ponds "just like" Dunwoody's and he knows it, regardless of how helpful he's trying to be.

"Then get in!"

"In" means into a blue pickup that belongs on a movie prop lot. The engine makes a loud, solid, engine-speed knock. I think back on all the times I'd visited salvage yards, wishing for family wealth so I could take a minimum wage job disassembling machines just like Dunwoody's truck. Images of Erector sets, Leggo blocks, toy airplane engines and bicycle rear axles, swirl through my head. We'd all like to become kids again; Duane Dunwoody's almost succeeded. Some would say a man who studies bugs has too.

We bounce through the back drive and onto a road covered, cathedral-like, with giant overhanging cottonwood branches. The bar ditch is an ode to poison ivy; red-headed woodpeckers hunker down on the sides of fenceposts as we thunder past; young rabbits dash through our dust into the roadside brush; a brown thrasher slips quickly from one shadow to the next. Dunwoody brakes hard, cranks on the wheel, lurches onto a double path nearly hidden in the grass, and slams to a stop. A string of coal cars, stretching as far as we can see to the east and west blocks our way. Duane Dunwoody is not a patient man and a string of Wyoming coal cars stopped in the middle of the Sandhills tries whatever patience he has this morning.

"Anyway, it's right over there." He points through a gap between the cars, wrestles the gears into reverse, pushes on the gas, sending the engine into a pounding fit of steel against steel. "Right to the right of that house."

"I'll call the owner."

The house is vacant; the owner lives and works in a small fishing community twenty miles away, on the north shore of a very large reservoir. In a week I'll explore the other pond, measuring its qualities against those of the most fertile, seething, and reliable sources of aquatic life, the small pothole microcosms of North American freshwater biology that dot the High Plains landscape. But even as we leave, I know that this new place will never match the richness of Dunwoody's. In a land where oases are far apart, and summertime is fleeting, you don't waste your time trying to study bugs where bugs are in short supply. That's why we come back, day after day, year after year, to a single small body of water constructed, maintained, and owned by one person.

Duane Dunwoody returns me to the field south of his barn, to the shore of his pond. The quiet gurgle of water falling into a pipe gives a sense of continuity to the obviously seasonal, short lived but repeated, elements of a Sandhills morning: tiny delicate flies, food for the swallows working the weedtops, equally tiny parasitic wasps, searching for victims among the seedheads, the warmth of the sun and stillness of a morning that, in late summer, could just as easily have been filled with primeval violence of lightning and hail. In the middle of a natural symphony whose score is dictated largely by the sun and weather, one body of water, Dunwoody Pond, is a contrivance, a blocked spring with a gurgling drainpipe.

The pipe leads underground for a quarter of a mile to the south where the overflow waters a hay meadow. After all the millenia, the shifting and marching of dunes, the births and deaths of species; after all the giant blowing clouds of volcanic ash, the comet and asteroid collisions, uncountable blizzards, gulley washers and frog stranglers, have made their marks on the Sandhills; after all the bison herds and plains Indians have passed their ghostly ways, taking what nature offered, moving on when the offers were withdrawn, Duane Dunwoody has come home to a place he ice skated as a kid, has dug out a low spot, put in a drain pipe, and in so doing bettered his odds of winning against every farmer's most feared enemy: drought. He's also created a water wonderworld. Furthermore, he knows it. But this particular pond didn't begin life as a device for watering a hay meadow. When he dug out the low spot and installed the pipe, Duane Dunwoody intended to raise catfish for food and profit.

"Shike-pokes ate 'em all!" he says. "Dug that pond, put in a thousand fingerlings. Ended up with only two of those fish!"

My grandfather called herons shike-pokes. Why, I wonder, out of all those available folk names—butterhump, mire drum, bonnet martyr, Lady of the Waters, fly-up-the-creek, quawk and qua-bird—would he pick the one my grandfather used? Grandfathers are people who bring back memories of childhood, of a raw curiosity fed by a knowing, gentle, but only partial, answer, of houses filled with old books, old books filled with pictures you can't seem to forget, and your parents smiling at the stories they once lived then heard told a thousand times. Now, on the shore of Dunwoody Pond listening to shike-poke tales, I might as well be back in that old wooden house with the dark woodwork, lying on the floor, totally consumed by a color plate of insect metamorphosis, with my parents calling from miles away—it's time to go home, Johnny.

"I used to come over here when I was a kid," he says. "Now anybody can come over here. Go fishing. You know, guys need a place they can take their kids fishing. Kids need a place to go fishing."

"I guess the shike-pokes do, too, Mr. Dunwoody."

"They sure ate 'em all. So I put in a bunch of bluegills."

From my experience, great blue herons will eat sunfish just as readily as catfish fingerlings, although now the water weeds are so grown up that maybe the bluegills have a place to hide from that stabbing bill. But nothing can hide from a stabbing curiosity. Duane Dunwoody might have failed as a fish farmer, but just by filling up a hole with water and letting nature fill up that water with whatever life could ride in on the prairie wind he's succeeded beyond anyone's wildest expectations as a grower of seductive puzzles.

Another truck, every bit as used as Dunwoody's, but bearing a wooden camper shell, is parked by the pond. An older man sits on the tailgate, a white bucket with a single fish in it at his feet. His companion is across the water, on the other side, studiously tossing a line, his silver goatee lending a profound sense of dignity to the traditional and common act of fishing. The two men could be great blue herons; they posses the same patience, sense of time and place, almost instinctive behavior that directs them to seek their sustenance along the shore.

I nod a greeting to the gentleman on the tailgate, step to the bank, and look down into another universe whose only familiar denizen is the fat sunfish guarding a nest in the shallows— a well-fed, aggressive, focused individual, its tail fin undulating subtly, nervously. Only the laws of chance could have dictated the fates of these two fish—the one guarding the nest and the one in the bucket. I'm surprised any human fisherman ever catches anything out of Dunwoody Pond; there is too much to eat, and the water is too clear, for a fish to get fooled by bait. The one in the bucket must have felt a rare hunger pang just as the deadly hooked worm drifted past. I envision a singular instant in which a chocolate chip cookie comes floating through the air within arm's reach. I grab with both hands, not seeing the committee assignment attached to the cookie by monofilament line. Poor dumb college professor, someone would then say, watching me swim in circles waiting to get fried and eaten.

As I walk the bank, vibrations zing through the ground and into the water. Each footfall sends a shock wave that jiggles nerve endings along the sunfish's sides. The animal responds, moving into the dark green depths, disappearing among the massive beds of *Ceratophyllum* whose stems and leaves make miles of bottle-brush boas, twisted and coiled back on themselves like giant chromosomes. There is a movement; the sunfish strikes. Death comes to another larval dragonfly, as swiftly and unemotionally as it has come to the hundreds of smaller scuds and wigglers consumed by the dragonfly over the past months. Sunlight pierces a small gap in the vegetation, a ray slices down into the weeds. Fine particles and diatoms drift in clouds throughout the light shaft; along a *Ceratophyllum* leaf, a busy microscopic, telescoping rotifer swirls a tiny current around its head, sucking in the diatoms; worms, smaller than bits of thread, pack the dense leaf tufts at the end of a stem, secreting mucous tubes, and pulsating, always pulsating, pushing, swallowing, pushing, swallowing, eating the drifting dirt from Dunwoody Pond.

Technically, we call this community of tiny plants and animals a food web. Technically, energy and carbon flow through such a community, the first trapped by algae and diatoms from the dim light that breaks through the eastern willows every morning, and the second caught from the foggy breath of a doe and her fawn that come to drink at sundown. These basic elements of life on Earth—energy and carbon—are then passed from one species to another by all the grabbing, gnawing, sucking, stabbing, swirling, licking, biting, chewing, rasping, and scraping that constantly occupy the times and lives of a prairie pothole fauna.

Standing in the grass along the bank, peering down into

the depths, I'm again reminded of the mind's power to construct a picture made only of abstractions. I don't actually see what I know, but know what I would see if I took a drop of this water and put it under a microscope lens. I don't actually see energy and carbon flow through a food web, but I know what I would discover if I did the right measurements—of birth and death rates, of numbers and biomasses, of radioactive carbon atoms accumulating in the various inhabitants of Dunwoody Pond. Beneath the lens I'd see an extraordinary array of life in constant motion. Upon the computer printouts of my analyzed measurements I'd see numbers and symbols, the kinds of abstractions that textbooks use to reconstruct still other abstractions—energy pyramids, stylized mayflies, fish, herons, mice, and hawks connected by criss-crossed arrows—that are supposed to tell the story of how Dunwoody Ponds everywhere work.

"This is science" claims some printed schedule, some declaration of content, some photocopied handout, referring to the diagrams and arrows. "This" is not science. "This"—the food web—is an artist's rendition, a sterile summary of decades of science—collections, experiments, identifications, struggles with arcane and obscure literature, hypothesis building and testing, failure, arguments in local bars, arguments in front of sophisticated audiences, rejected grant proposals, manuscripts submitted for publication, theses written, advanced degrees awarded, and publications. All this activity rests on ideas, ideas that spring from the heads of curious people seeking explanations for complex mysteries such as "what role does a microscopic worm play in the greater scheme of things?" and ideas that come from the brains of people gone fishing, like the man on the tailgate, in places like Dunwoody Pond. The textbook food web diagram is missing its top predators, namely the human beings who nourish their minds off the tiers of reproduction and predation descending into the infinitesimal realms.

Hot morning sunshine melts my sharper inclinations, turning a coldly neutral scientist's mindset into an aesthetic experience, a symphony of insect sounds played in an art gallery whose floor is grass. The knowledge that a puzzle can be solved recedes into the soft inner depths of metaphysics, a field of questions that begins with "why" instead of "how" but ends with the words *Dunwoody Pond*. "How do I study the food web of Dunwoody Pond?" is a very different question than "Why do I study the food web of Dunwoody Pond?" You can find the answers to the first one in books, and although you'll find more than one answer to the How question, you won't find an unlimited number of such answers. But you can't find the answers to the second— the Why question—in a book; instead, you must search for them in the fields south of Duane Dunwoody's barn. And yes there are a great many such answers, far more than to the How question. The number must run into the hundreds of millions, possibly even as many answers as there are human beings who could be enticed into seeking them. You think, if only these people had access to Dunwoody Pond, and the luxury of wonder time to go along with their labor time—what a different world we would live in. That world would contain the hundreds of millions who could easily distinguish between what some authority told them and what they knew they'd seen, or not seen. In the final analysis, this conflict between declaration and observation provides the electricity that drives a scientist. But when authority is ignorant, and the scientist knows it, and is young, then that electricity becomes a powerful surging current.

Throughout recorded history, scientists have routinely been perceived as magicians, pulling discoveries off the lab bench, holding them high as if to say See! A new chemical compound! A new bacterium! A new theory! How do magicians do that you wonder, as the rabbit hops into the waiting basket, the white dove flutters over to the shoulders of a glamorous stage assistant, the brightly colored flowing scarves appear, one after another, from thin air. The knowledge that this stage magic is of human origin makes it all the more fascinating as entertainment. And the knowledge that science has a mystical side makes it more captivating, I sometimes think, than other forms of adult entertainment.

But out on the intellectual streets, those who search for underlying mechanisms are often considered the scientific nobility, while the magicians, those who appreciate the show, and understand its role in providing society's scientific and technical human resources, are sometimes portrayed as charlatans. In the legislative hearing rooms, grant proposal review panel meetings, foundation boardrooms, symposia on technology and economics, we tend to tell ourselves that a scientist's job is to decipher the magic, not perpetuate it, to explain the wondrous, not compound it, and ultimately, to deliver the power of an explained mystery to those who would wield power over human lives. All such arrogance dissolves on the shore of Dunwoody Pond. The mystery of the universe, written small and near, surrounds you, crawls up your leg, flies into your ear, bites you on the neck, and leaves black muck under your toenails. You are suddenly alone, surrounded by items you don't recognize and have no practical use for. What am I doing here? you ask—the ultimate human question rephrased in highly local terms.

In the history of humanity's efforts to explain the universe, the question seekers ultimately have made more of an impact on our daily lives, as well as on our cosmological lives, than have the question answerers and technocrats. Those who search for problems have found plenty to keep the rest of us occupied. The problem seekers are the real magicians, the ones who seemingly pull illusion out of thin air. But in fact problems do not come out of thin air, nor are they only an illusion. Instead, they are works of art. They originate inside a human mind; they are made of experience and curiosity mixed in with a dash—sometimes a sizable dash—of naiveté and aesthetics, the building blocks of wonder. And so we come back to Dunwoody Pond, week after week, year after year, digging and netting for questions, which we put into white buckets made of recycled plastic, then carry around with us for the rest of our lives.

Up on the shore, from the tailgate of his truck, the fisherman I'd nodded to pulls in his line, jerking at the hook caught in the mass of aquatic plants, finally dragging a strand up into the sandburrs. A bloated worm still sags, wrapped in Norman Rockwell fashion, from the curved steel shaft.

"Too much to eat down there, Titus," I offer an explanation for his lack of success. "No bait in the world looks good to 'em." I point into the bucket. "This one's just doing you a favor. Figured you needed some encouragement. Or maybe he's just the dumbest fish there is. But I think his friends sent him up here to do you a favor, make you feel good."

Titus laughs.

"I know it!" he answers. "Probably ought to just look at 'im for a while, then toss 'im back."

Titus doesn't care about catching fish; all he cares about is being where he is, enjoying the sunlight, the sound of water falling into the pipe, damselflies in the foxtail, a hawk high above the cottonwoods, and the smell of prairie. He's fishing for peace instead of questions, a peace of mind and body, an assurance that beneath that surface film of violence and conflict, the world is orderly, predictable, delicate, elegant, neutral. But like all fishermen everywhere, Titus knows about parasites.

"Little white things on the gills," he says, describing *Ergasilus,* a crustacean that hangs on with its antennae. I tell him what they are, how they get onto the fish, and guarantee that he cannot be infected. What I can't guarantee, however, is that he won't get infected with the social diseases you pick up by messing around in Dunwoody Pond: virulent new attitudes, proliferating new perspectives, and thoughts about changing your life's directions, all such thoughts and attitudes potentially as parasitic as the crustaceans on a sunfish gill. Get heavily infected with new thoughts and your behavior might get altered. You might find yourself suddenly changing directions. Titus knows about changing directions; it happened on his first visit to Dunwoody's.

"I came out here in seventy-six," he says, "wanted to be a cowboy. Bought some brand-new jeans. First day on the job, got those new jeans caught in the power takeoff." He knocks on his knee; the rap of knuckles on wood pings through the morning air. "Tore my leg right off."

"And you're still here."

"Right here. Not goin' anywhere."

"You might be in the best place on Earth, Titus."

"I believe that."

Power takeoff is a shaft that spins, engine speed, out the back of a tractor. This spinning shaft is used to run grain augers, pumps, and a variety of other machinery, but you have to hook up to it. And that's when disaster can strike with stunning speed. Everyone who lives on the land has machinery horror stories—lost fingers and thumbs, caught clothing and hair, the shocking instants in which routine work suddenly becomes tragedy.

The ping of Titus' wooden leg is a stark reminder that the isolated rural houses of America are occupied by people who work at one of the most dangerous of occupations: their tractors, bailers, mowers, storage bins, propane and gasoline tanks, large and strong animals; the weather they talk about incessantly because it dictates the form of their daily activities; the isolation itself, the distance from emergency care; the years of sun and windburn; the creeping effects of age and its erosion of their strength all make a mark. Over the years, the marks add up. Old farmers and ranchers are survivors in the strongest sense of the word. But if you ask any of them where they'd rather be, the answer is: back home, back amid the risks, standing in their own pasture at daybreak, washing the dust off their faces before dinner, sitting at an antique table, saying grace, and digging a fork into homemade rhubarb pie.

Duane Dunwoody, for all his mental worldliness, is a perfect case study. Born and raised in the tiny village of Keystone, a mile south of his pond, he was not easily dislodged, staying behind when his parents moved a hundred miles west.

"I was an ornery kid," he offers when asked about his past. He's still an ornery kid, I think sometimes, listening to him evaluate candidates for public office, pending legislation, new cars, or railroad and utility companies. A stint in the U.S. Navy took him to San Francisco, where he later went to college. After sixteen years as an elementary school teacher in Santa Cruz, he came back to the same ranch where he played and worked as a child, made his pond, fed his investment to the shike-pokes, gave Titus a place to live without a leg, and piece by piece, calf by calf, haystack by haystack, built his little empire by always coming back to the land that was just beneath his feet. One man's labor builds a house of the mind for those who cross his path. I've seen it happen many times—the personal dream turned into society's treasure.

Maybe this phenomenon, repeated so often with respect to our great museums, galleries, zoos, libraries, is what drives a teacher, too. Maybe the presence of Dunwoody ponds allows us to flex the idealism which seems to be the sole property of philanthropists, namely the construction of an edifice that provides a sort of binding cement to the human experience. But our idealism is of the highest and most ethereal kind, compared to that of the big donors, for we build castles made of ideas, and these ideas are stacked one upon another inside the heads of our students, and only after the decades have passed do we get to see our efforts brought to fruition. We know how to put the castles together, however. All we need is a source of exotic puzzles. A Dunwoody Pond. Then we lead a young mind to the edge, give the kid a net and a microscope, and get the hell out of the way. So on this morning, as is a teacher's habit, I've delivered another victim—a young woman named Aris—to Dunwoody's, and as Titus and I discuss the relative intelligence of various fish, and their consequent relative ease of being caught, she steps into the pond, carrying a large screen cage with one hand, and some wooden stakes and wire with another.

Like Titus' sunfish, Aris too has been caught by something she saw in Dunwoody Pond, hooked into an obligation that will likely never end, and has returned in order to entangle herself still further, and deeper, by fishing for abstractions. She was caught late one night while she sat alone, her eyes pushed up to the microscope lenses, a bucket of Dunwoody mud, water, and vegetation at her feet, and a table full of exotic glassware spread before her. Strange music issued from the "box" on the shelf above her head. Fish tanks gurgled from the shelves behind her back. There in the strange little room, she saw a thing of great beauty—a single cell. It was a very large single cell, the largest among a dozen or so similar ones, and it had a structure of elegant complexity at one end—a crown of thorns—hooks, actually, radiating out in six pairs. The cell was divided by a fine septum into two unequal compartments. The cytoplasm was dense and granulated. And as Aris watched, the cell moved slightly, a slow gliding derived from no visible means of locomotion. When she put the microscope on a higher power to study the cat-claw hooks, she saw that they were at the end of a fluted stalk. As she scanned the preparation, she then saw other kinds of cells, with complex structures something like a lampshade instead of the crown of hooks. She studied these cells for an hour, memorizing their features, absorbing the information they conveyed to her. And at the end of this time she said, "This is what I am going to study."

She had no idea why she was going to study these cells, only that she wanted to, and because she wanted to, would find a way to do it.

"It's a bad idea," I said, one of the few times I've done that to a student who's made such a decision. "This system is so unpredictable, so uncontrollable; damselflies are pretty uncooperative in the lab; the smaller naiads are impossible to identify; you're three hundred miles from home. There are a thousand good reasons not to pursue a bad idea."

"You can't tell if an idea's bad until you pursue it." She defended her sudden interest and returned to the microscope.

The cells she'd found were actually parasites inside the intestine of a nymphal damselfly, the immature stage—also called a naiad—that lives for months beneath the water before climbing out onto some emergent vegetation, shedding the nymphal skin, stretching its wings, and, in a little while, flying, usually far from its home pond. Damselflies held a special attraction for Aris, perhaps because of their delicacy and beauty, or perhaps because of their similarly special, and serious, attraction for her friend and lab-mate Tami, who was also studying some of these parasites, but in the adult insect. When a damselfly undergoes its final molt, into an adult, its habitat changes from fully aquatic, living and feeding beneath the water, to fully terrestrial, living, feeding, mating, on land, usually near some body of water where it will drop its eggs, and dying, often eaten by a swallow or a spider. Tami's story—her eventual choice of damselflies—is worth telling in its own right, and so you'll get to meet Tami later, on her own journey around the shores of Dunwoody Pond. What separates the two friends at the moment, however, is their respective focus on the insect's two life-cycle phases. Although the adult is the same species as the naiad, the adult lives in a totally different world from that occupied by its earlier self.

Now here is the mystery that captures the attention of these two young women: During its metamorphosis, and with its emergence, the damselfly loses its parasites. But sometime later on—nobody knows when or how—the adult gets its parasites back. In violation of the Homerian advice, repeated by the Apostle Paul in his correspondence to the Corinthians, still read daily at weddings throughout the Western World, damselflies do not give up their childish ways entirely. They may leave the childhood environment, leave behind their leaf-like childhood gills at the end of their abdomen, leave behind even their skin—a ghostlike shadow of themselves as children—and take on adult wings, but eventually—like grownups who study bugs—they get back their childhood parasites.

The immature, or aquatic nymphal, damselflies also have a property that the adults do not: a succession of molts, sizes, enemies, food choice items, and probably microhabitats. During its growth from a hatchling into a fully developed naiad ready to metamorphose, the insect increases in length, stepwise with every molt. The last immature stage may be ten times the size of the one that left the egg, sometimes a year earlier. But during its aquatic life, the naiad usually acquires parasites, occasionally several species, often dozens of large individuals you'd think would clog its intestine, and all with wondrously complex anterior ends that sink deeply into the layer of cells that lines the damselfly's gut.

The intestinal lining is battered by disruptive forces. A damselfly's gut is lined with enzyme-filled cells that get torn off by the sharp, angled insects and crustaceans the damselfly eats, prey items that end up looking like miniature versions of crushed automobiles in a salvage yard. When the gut cells are torn away, the digestive enzymes are released. Cells are replenished from pockets of tissue lying below the lining, in the muscular wall of the intestine. The parasitic cells that Aris has chosen to study live in this turbulent and dynamic milieu, and may well have lived there, as species, for unimaginable lengths of time. Damselflies have been on Earth since the Carboniferous—three hundred million years. There is every reason to suspect that one-celled animals have occupied the intestines of various arthropods for at least that long. Fossil amoebas tell us that had we walked the Ordovician shores with a hand lens, looking closely at handfuls of sand between the tide lines, we'd see familiar foraminiferans—those ubiquitous inhabitants of the ocean whose shells provide a continuous record of evolution, extinction, and microscopic diversity over the past half billion years. Annelids, considered by some to be ancient relatives of arthropods, possess their own internal fauna of one-celled animals. Aris goes to the museum and finds a Devonian starfish, more proof that truly ancient life forms are still present on the planet and, furthermore, proof that even a child can understand. Her wanderings have prepared her for certain interpretations of the world she observes. So when she goes to the lab, dissects a damselfly, finds an entire community of parasites, she must believe, because she knows how long damselflies have been on Earth, that she's seeing a relationship of equal age.

By the time she waded into Dunwoody Pond with her screen box, Aris had been cutting open damselfly naiads, counting, measuring, and separating intestinal fauna according to kinds, and analyzing the naiad molting process, for half a summer. She was convinced that a progression of parasite species marched through the Dunwoody population of damselfly naiads, those present in June being replaced by others in July and still others in August. An explanation for this phenomenon seems a problem worthy of her thoughts for the next two years. The screen box is a first step on her intellectual journey; the box is an enclosure to hold damselfly naiads, of a certain size class, containing a particular mix of parasites. Back in her laboratory, Aris has a sample of this "cohort," part of a supply of naiads she'd collected earlier and dissected. At the end of the summer, she'll return to Dunwoody Pond, retrieve her screen box, recover the naiads, and discover whether they've grown the same amount, and gathered the same array of parasites, as the ones she has back in the laboratory.

In essence, in addition to her specific biological questions, Aris also is asking whether the research she's doing in the lab has any relevance whatsoever to the events that occur in nature.

This aspect of her quest enhances the metaphorical qualities of it. She could be asking the same question at a higher level, on a larger scale, both of herself and of the society in which she lives. The subject of her study—the seasonal progression of species through a habitat—and the explanation for such a regular change, are both intangible entities. You can't hold a progression in your hand; you have to watch it happen. And explanations are equally ethereal; the most uncompromising scientists consider all explanations simply assertions to be tested, not hard and fast rules of cause and effect. The only tangible product of Aris' enterprise is herself—the well educated person, the human being who seeks explanations in her own personal observations, counts and analyses, rather than in the myths, claims, and declarations, of others. Does what she's doing in the lab, i.e., the "experiments" intended to turn herself into a scientist, have any relevance whatsoever to the events that occur in society? Is there any "need" for well educated people in her society, people who seek answers in analyses rather than in myths? Yes, she concludes, wading out of Dunwoody Pond, turning to look at her screen box staked down below the water's surface. Yes, there is a need for my kind. And in order to produce my kind, she concludes, I will have to come back to the place where my idea lies just below the surface, held down with stakes and wire, in there among the *Ceratophyllum*, the rotifers, the sunfish, and all that snapping and eating and breeding.

Over the next few weeks, Aris returns to Dunwoody's over and over again. Each time she looks at the screen box, lying just below the surface. Sometimes Titus asks me about what she might find when she retrieves the box and opens it; you can tell he's been watching it, wondering what, if anything, might happen. I explain the experiment as best as I can. Titus nods and smiles; he understands completely what it's like to decide you're going to be something—a cowboy, a scientist, then actually, physically, carry out the decision—buy new jeans, or put a screen box in Dunwoody Pond. You can tell that Duane Dunwoody's also been watching the box. I assure him it will come out by the end of the summer. He says don't worry, Doc. He means the box isn't going anywhere; you get the feeling he agrees that kids need water, mud, and bugs in order to get a proper education. And maybe fish, too, but certainly the water, mud, and bugs. And after a month has passed, Aris drives out to Dunwoody's, wades into the water to retrieve her bugs and mud. Algae covers the bottom and sides. She opens the door and washes the contents out into a bucket.

"They survived," she says, "and grew."

I'm not surprised. I've come to associate survival and growth with Dunwoody Pond. But did the damselflies in the lab do as well? Sort of. No wild thing does as well in the lab as in nature, but these came close. Of course Aris had put some of Dunwoody's *Ceratophyllum* in the lab aquarium. Maybe the ones in the lab smelled their home pond; maybe that's the reason they did reasonably well; maybe they thought they were back at Dunwoody's. Then again, perhaps damselflies were more tractable than Aris at first thought they might be. She's willing to accept that possibility. If true, it will make her life much easier than she'd expected it to be, over the next two years.

At summer's end, Aris takes Dunwoody Pond home with her, literally as well as symbolically. She fills up several five-gallon buckets with mud, water, algae, *Ceratophyllum* and loads them into a van. Then she runs an air hose from a battery-operated pump through a hole in the lid. The reassuring purr of these pumps is a sound she'll hear for the next seven hours. Before she's out of the barnyard, water sloshes out of the lid holes and she gets a whiff of Dunwoody's; the smell is as reassuring as the pump purr. Three hundred miles east, in a concrete lab with bare ceiling pipes and ductwork, she dumps each bucket in an aquarium, runs an air line through the hole in the bottom of a clay flower pot, puts the pot in the aquarium, and starts a slow but steady stream of bubbles. There. Aris' Dunwoody Pond.

Outside the leaves turn brown and fall; massive crowds on their way to ballgames surge past the building where three little Dunwoody Ponds sit bubbling; surly clouds blow past the fourth floor window from which a cold light falls on the aquaria; the clouds spit sleet, then snow, then ice, and all the world, it seems, is frozen, suspended in time. But inside, beneath fluorescent lights, the little Dunwoody Ponds boil away. Tiny midges emerge from pupae, fly up to the lights, then die, accumulating on top of the long bulbs and along the window sill. Ostracods, almost microscopic crustaceans that look like fat swimming clams, dump hundreds of baby ostracods among the plants. The babies join the parents then, in swarms that gather just below the surface. Planarians—flatworms—course along the glass, randomly gliding, waiting, it seems, for a fallen snail, bird, or fish. Periodically the planarians get a chunk of chicken liver, which they attack *en masse*, forming a seething blob of pulsating worm until, satiated, one by one they break out of the mass and return to the glass walls, sluggish, their Y-shaped guts, packed with blood-red liver, showing through their mud-tan bodies. In confinement, the snails' populations explode; dotted jelly masses of eggs appear on the glass; the dots get larger; then the young snails emerge. Eventually the snails attack the chicken liver, too. Several times a day, humans walk past the aquaria, pick up a hand lens, then sit down on a nearby chair to study all the activity in the water behind the glass. When the humans walk away, they always smile and look back. Sometimes they think: My how those worms have grown!

An aquarium filled with colorful gravel and tropical fish is a very different contrivance than one filled with plants, mud, worms and insects. The former often seems more of a statement about the aquarium owner's pride than about the inhabitants; the latter always tells you more about the "owner's" mind than about his or her pride, and tells you more about the inhabitants than about the owner. In fact, the aquarium filled with mud, worms, and insects can't be owned; it's only borrowed. Nevertheless, there's something reassuring about being able to visit Dunwoody's whenever you want to, putting the rest of the world aside for a few moments while you enter that microscopic realm, mentally, through your hand magnifier.

Then one day in April, a cold, rainy day at the end of a similarly miserable week, Aris unlocks her lab door and does what she does every day upon entering the room filled with equipment, tanks, fluorescent lights, microscopes, and computers—she walks over to the aquaria. A damselfly has emerged during the night. It rests on the edge of the glass. Almost instinctively, Aris puts her finger up to the insect. In her mind, the damselfly crawls up on her finger. Instead, at her touch, it flies to the window. Yes, thinks Aris, one of my two years has passed; it's about time to return to Dunwoody Pond. Rain pounds against the window. The damselfly flutters along the sill. Summer's two months and three hundred miles away.

2
Choosing Damsels

And I serve the fairy queen,
To dew her orbs upon the green.
— A FAIRY

Scientific names remind me of foreign diplomats, suddenly cast into the light by events half a world away. Duane Dunwoody hears odd voices on television and accepts them as a necessary element of his now global communications network. Other Sandhills families, even more physically isolated than Dunwoody, do the same. So if we are to sit around the dinner table and talk about political forces ripping at the human fabric, we must mouth unfamiliar words. And if we're to talk about delicate beauty, frailty battling the prairie gales, striking microscopic colors emerging from a vile and smelly froth, Paleozoic patterns now resting on our outstretched finger, we must also make our peace with ancient languages spoken in exotic places.

Tami handles such words easily; she's practiced them daily for years.

"*Ischnura verticalis*," she says, easily and smoothly, with the softness of someone recognizing a tiny friend in a far off land. She slips *Ischnura verticalis* into a hole in the lid of a plastic gallon jug.

"*Enallagma civile*," she calls the next one, just as gently and easily, just as instructively, and slips *Enallagma civile* into the same hole where *Ischnura verticalis* disappeared. Tami is choosing damselflies, an activity in which she can become totally and completely absorbed in a world of her own, progressing slowly through the weeds.

She flicks the net. A soft whisper of gauze brushing grasstop floats across the glassy surface of Dunwoody Pond. She holds the fine white linen up to the sunlight. Inside a pair of damselflies flutters against the webbing. Their membranous wings sparkle in the glare, sending iridescent flashes through the cloth. Tami reaches down into the flimsy bag, carefully working her hand through the fold until her fingers press gently on the wings. Death awaits *Ischnura verticalis*, for Tami is a businesslike reaper. And as surely as she's chosen one *I. verticalis* out of the thousands that rest, chase tiny prey, and seek mates along the shore of a pond, she's also chosen a path into the next century, a path aligned closely to the fates of her insects and the other animals that live inside them.

Her trek through the arcane jungles of Invertebratology began when she was given a small card with another odd name on it: *Siphonia tulipa*. Go to wonderland, she was told, and find *Siphonia tulipa*. But when she climbed the shining marble staircase and pushed open the ancient creaking doors, she found so many elegant items that she forgot *Siphonia tulipa* for a time, and became lost among the rock leaves, stared back at the stone eyes looking up at her from their beds of green felt, took a trip back four hundred million years, riding there in the frozen writhing arms of a black star, felt sadness for the crushed flowers that were not real flowers at all, but sea lilies, from a far off time.

Around her feet the children played and ran calling to one another to come look at all the strange creatures made of rocks and epoxy and information and the hard work of people who dug into the Earth for evidence of past worlds. I must find *Siphonia tulipa*, Tami thought, eventually, and when I do, it will be the most beautiful of all these wonders. She was wrong. It was not the most beautiful nor the most complex of fossils in the museum, but it was hers, for upon the card she'd been given was not only a lyrical name, but also an assignment: Write a story, about *Siphonia tulipa* that will make one of these children want to grow up to be just like me—forever young of mind, forever curious about the lives I cannot live. She leaned over then, staring closely through the glass, and asked her questions of the rock: What is your secret? How do I make a person choose an animal, then because of that choice, choose a life, just by telling a story? What kind of a story might this one be?

Now, in the hot midmorning, Tami stalks through the weeds with the same sharp curiosity that she'd had entering the museum. She's just as surrounded by wonderland, just as aware of her ultimate task and the labor that follows, and just as ready to ask the same questions of *Ischnura verticalis* and *Enallagma civile* as she'd asked of the cold rock *Siphonia tulipa*: What is your secret? But she's older now, five years down the road, and she knows the secret: Giant problems have giant powers of attraction. They consume your thinking time, lead you into exotic and dangerous places, turn you into a monster your friends don't recognize. So you get new friends, people who walk through the weeds and choose giant problems as easily as they choose *Ischnura verticalis*.

And she knows, too, the kind of story she has to tell in order to make some child choose a damselfly, or a fossil, or for that matter a beetle, a worm, or a bird, as a guide to wonderland.

The story cannot have an end, only a beginning, then a middle with an infinite, branching, interconnected maze of pathways. When the child enters the maze, expecting to find an answer, an end, she sees only choices, and these in turn are never clearly defined as right or wrong. Ahead lie many roads, all disguised as something they are not, all leading into scenery that can shatter your perceptions of a well-organized universe. I was that child once, Tami thinks back, and smiles at the memory of a card with a magic name: *Siphonia tulipa*. And I wrote my story, and it did make one kid want to become a scientist. I'm that kid! She flicks her net, choosing damselflies on the shore of Dunwoody Pond.

But Tami's new tulip does not lie frozen in stone in a museum case. Within her chosen damselflies lives an astonishing array of other animals. In the laboratory, she peels open an intestine, using her fine forceps to tear a strip down one side, causing the rest of the tube to turn itself inside out. Sometimes a dozen long white bodies then appear, their "heads" buried into the gut wall, between the cells. These are the parasites that Tami has picked for the topic of her mental labor. If she uses them properly, they will open doors for her, carry her to a podium in a far off city where she'll throw her ideas out for discussion to an auditorium full of scientists waiting to see how well she succeeds as a member of their club. And no matter what happens to her for the rest of her life on Earth, these odd, elongate cells will sit beside *Siphonia tulipa* in her memory as the pieces of nature she used to build her career.

But even as she cuts the tiny damselfly intestines, and gently teases the parasites away from the gut lining, Tami knows that she faces two tasks disguised as one. Her ultimate goal is to reveal the various mechanisms by which these one-celled animals attach to the damselfly intestine. Before she gets to that point, however, she must deal with several tongue-twister names. *Siphonia tulipa* was lyrical enough to make her want to say the words; *Ischnura verticalis* had a certain mixture of hard and soft sounds, like chocolate and salt, that she enjoyed; *Enallagma civile* reminded her of a relatively tame, but nevertheless entertaining, jigsaw puzzle. Except for one species, however, the long one-celled parasites in the damselfly intestines have no names. Instead of memorizing exotic words, Tami must make some up, then defend her reasons for assigning them.

"This is the one I'm going to call *dunwoodii*," she says, leaning back from the microscope. Duane Dunwoody, for all his bluster, has been a friend whose help cannot be measured, nor adequately priced, nor even repaid, except in the most respectful and quiet way: an honorific name, published in a scientific journal, thus spoken forever, around the world, whenever anyone talks about the animals that live inside damselflies that carpet the shore of Dunwoody's.

In naming parasites after people who've been a significant part of her life, Tami follows in the tracks of another young woman, Sarah, who also came into the Nebraska Sandhills to study biology and ended up naming the one species of damselfly parasite Tami recognizes.

Sarah walked into my lab one day and said: "I'm here to do research, but it has to be on something nobody else has ever worked on." At the time, Sarah was an undergraduate at Brown University looking for field experience out on the western Great Plains.

"Study the parasites of damselflies," I suggested, "if you really want to work on something that nobody else has studied."

"What kind of parasites?" she asked.

"They're called 'gregarines,' " I replied, "and they're the most insignificant, unappreciated, mysterious, and economically unimportant animals I know. Which is probably why nobody else has studied the ones in damselflies, at least around here. But," I added, "they're beautiful, too, and reasonably captivating."

"How do I start?" wondered Sarah.

"Write fifty questions," I answered.

"About animals I've never seen?" She was getting the picture quickly—a positive sign!

"Well, go find some, then write your fifty questions," I said, handing her an insect net.

Sarah, like Tami in the museum, went searching for her animals, which she found by the thousands along the shore of a place called Martin Bay Pond, about two miles away from Dunwoody. Martin Bay Pond has long since dried up; Sarah's two summers along its shore may have been a singularity—an idyllic, highly instructional, emotionally captivating, intense intellectual experience which can never be repeated because the place where she had it has disappeared. In fact, the disappearance of Martin Bay Pond stimulated the search that led, eventually, to Dunwoody's. We'll come back to Martin Bay Pond, or rather to the dried mud bed, later, when it's time to talk of droughts, both natural droughts which lay bare the land and bring into prominence the hardiest, oldest, and most tolerant of organisms, and droughts made of bad human decisions which lay bare the lands of opportunity and bring the most creative, resourceful, individualistic, and blasphemous minds out into the open where they flourish. But at the time of Sarah's explorations, Martin Bay Pond is full, and damselflies—*Enallagma civile*—blanket the tops of grasses growing right up to the water's edge. With one sweep of her net, Sarah is able to get a week's work. The first damselfly she cuts open has nearly three hundred parasites.

"What are they?" asks Sarah.

"Probably members of the genus *Actinocephalus*," I reply, looking through the microscope, "but you have to discover what their spores look like before you can be sure."

"What's the species?"

"I don't know. There's a lot of literature to consult, a lot of measurements to make. Maybe you have a new species. Maybe you'll have to publish a description."

A strange smile comes over Sarah's face.

"That would be great," she says, "before I came out here, I told my sister I was going to name a parasite after her."

I don't pursue that line of conversation very far. The night she discovers the spores, we celebrate with microwave burritos from Pro-Mart, the twenty-four-hour-a-day filling station that serves as the emergency ration source when discoveries that need celebrating are made in the middle of the night. The name of her undescribed species of *Actinocephalus* had been decided before she'd caught her first insect. Her sister was about to become immortalized in print.

"Carri Lynn, have I got a present for you!" says Sarah, taking another bite of her microwaved burrito. After two years of dissection, counting, measurement, and library research, she's convinced her species is a new one and is ready to write her description.

Sarah's two summers studying the parasites of damselflies are of significant help to Tami. Out of the five or six species of large gregarines in damselflies along the shores of Dunwoody Pond, only one, *Actinocephalus carrilynnae*, is familiar and identifiable. Sarah has gone to Arizona to pursue other questions, but her contribution stays behind to help those who follow in her steps. *Actinocephalus carrilynnae* is one of Tami's species whose status is, at least temporarily, defined and accepted. The rest of these species constitute, as is sometimes said in the profession, a "taxonomic mess."

The first time students encounter such a mess, right under their noses in some common and familiar place, the edifice of scientific knowledge suddenly appears cracked, if not shattered. Few experiences point so sharply, so quickly, and in such easily understood terms, toward that vast sea of ignorance every practicing scientist knows is "out there," as being unable to identify an animal using available literature. Parasites inside small animals are particularly unstudied. Of all the millions of damselflies that scientists have watched, collected, put away in museums, relatively few have been examined for parasites. And of those that have been dissected by someone looking for parasites, three young women from the prairies—Sarah, Aris, and Tami—are quickly accumulating the world's overwhelming majority. Not surprisingly, the first thing they discover is that they are suddenly among the world's experts. The second thing they discover is that in order to answer any question of process, they must first answer the question that plagues all ecologists at some time in their careers: What is it?

The gregarine parasites of damselflies are relatively large, for single cells, and their differences are manifested primarily in two features: their anterior ends, with which they "hold" onto the damselfly intestine, and their spores, more properly called "oocysts," by means of which they get distributed throughout nature. Tami has decided to focus on the first of these features, the holdfast structures, although in a larger sense, she's actually studying evolutionary events that probably took place a hundred million years ago between two species' cell membranes. Tami's convinced that with the electron microscope, she can see differences between species' solutions to a common problem: how to hold on to your place and complete your life history in a turbulent, mushy, environment. The unspoken assumption is that if she discovers how various species accomplish this daunting task, then maybe she'll be able to apply such knowledge to her own life history, hopefully to be lived out in academia—a no less turbulent or mushy environment than one finds in a damselfly gut.

The names of her animals, however, remain in her head instead of on the journal page where they'd be of use, and furthermore, the names themselves are neither fully formed nor firmly affixed to their respective parasites. Back in the city, Tami consults another lab mate, Rich, who's chosen beetles as his source of mystery, reputation, and career. Beetles, as a group may be the most common animals on Earth; there are a quarter of a million described species. Most of these species contain their own microfauna, and whereas the parasites of beetles are better known than those of damselflies, still only a handful of parasitologists have looked inside beetle guts. When these scientists have published their works, the papers have often appeared in old, odd, and foreign journals that few American libraries contain. Through patience and diligence, Rich has accumulated file drawers of obscure and convoluted literature, as well as a lexicon of scientific Greek and Latin words and their meanings. Together, microscopic animals in one hand, this literature in the other, the two young scientists search for the perfect syllables.

Euphonious and descriptive are Rich's personal criteria for names; euphonious and published so she can get on with her work are Tami's. Their search for perfect words reminds me of a Michael Lipman story—*The Chatterlings in Wordland*—that is among the treasured items remaining from my childhood bookshelves. The Chatterlings were delightful little elves dressed in red-tailed jackets and pointed caps with a pair of feathers. Their eyes were mostly white circles, i.e., wide open. The King informs Prince Tip o' Tongue that he's ready to retire and turn the kingdom over to Tip, but the Prince has to get himself a crown. When the poor kid comes back from the Royal Hat Maker, he has only a coronet. Of course as punishment, Tip gets sent on a rambling search, not unlike Tami with *Siphonia tulipa*, for the pair of words that means exactly the same thing.

The Chatterling-type search also leads to a history lesson. Among the obscure scientists who'd cut open damselflies looking for parasites is an Izushi High School teacher, Kinichiro Obata. Obata published descriptions of many species of parasites from Japanese insects. Tami and Rich know these insects; they've been at the microscope themselves, fine forceps in hand, pulling out an intestine from some of the same species Obata studied.

Obata's published paper, however, contains a narrative that the young scientists hope never to have to write in one of their own:

> I began the study of gregarines of insects in 1942, but I lost many data and manuscripts by the fire caused by the atomic bomb dropped on Hiroshima. After the second World War, I came back to my work, and ressumed [*sic*] some parts of my previous study.

Although the collections and manuscripts may have been lost, Kinichiro Obata's papers tell us that damselflies, and their parasites, as well as a person who studied them, survived a nuclear weapons attack. None of Obata's assigned species' names commemorate the war. He names a parasite species *tokonoi* because its insect hosts were collected near Mt. Tokono; another he calls *ozakii* and dedicates it to his "respectable professor," Dr. Y. Ozaki. Others he names after prominent physical features, a decision that Tami and Rich are somewhat inclined to repeat.

"*Steganorhynchus* doesn't sound quite right," I offer my opinion on the pair's choice of complex words for the generic name. "That sounds like a dinosaur instead of a parasite." The animal Tami proposes to name after Duane Dunwoody not only is a new species, it's also a new genus. If the descriptive paper is accepted by a scientific journal, the animal would be known as *Steganorhynchus dunwoodii*.

Stegano- translates into "sheathed" or "covered," rather like a lampshade; *rhynchus* translates into "nose." The *Stegano*- describes a delicate, membranous veil that adorns the end of this parasite's attachment stalk. So to honor her local rancher, Tami picks a name that means Dunwoody's Lampshadenose. What is a Dunwoody's Lampshadenose? A one-celled animal with a long stalk at the front end and a membranous, lampshade-like struc-

the most senior of these scientists commented in private on the paper involving insect parasites. For a mixed audience, he said, the student should simplify his jargon, and especially so if he [meaning the student] gives the same material in a job interview seminar. I pursued the meaning of the term "jargon," having heard plenty of jargon that nobody ever bothered explaining to me being presented as cutting edge science. You know, he said, all those names. Maybe he should just call them parasites *A*, *B*, and *C*. The problem, it seemed, was in the scientific names used by this student in front of an audience made up of biologists. How, I wondered, can you speak to professional biologists if you can't use scientific names? Then it dawned on me: The Latinized names of plants and animals are almost symbols for nineteenth-century biology—the Golden Age of Exploration. Nobody talks that way any more unless, of course, they want to make sure everyone else knows exactly what kinds of organisms are being discussed and have some sense of the evolutionary histories and relationships involved in the discussion.

At the other end of the spectrum of scientific education are the well educated professionals—usually businessmen and their wives—who say to me, at various social functions, comments like "John, I really couldn't make it through your last book. I hope you understand. It was just so difficult and so technical." Difficult and technical? I work so hard to make them easy and non-technical. Compared to the junk bond, savings and loan, and terrorist financing scandals of recent years, all reported extensively in local newspapers, the life of a parasite is relatively simple and straightforward. I usually tell my friends that. Then they get apologetic. Oh you know, they say, all those complicated names. The scientific names? I ask. Yes. Then I wonder if maybe their parents should have bought them a copy of *The Chatterlings in Wordland* when they were children.

But I'm usually polite enough not to express that wondering out loud. Instead, I make a comment about my class roster. Every time I record grades for two hundred students, I relive the colonization of America, the survival through three and four generations of Eastern European, German, Scandinavian, and Irish names, now carried with pride and a sense of cultural continuity. But recently my class rosters have carried other kinds of names, too—Hispanic, Indian, Pakistani, Asian, and especially Vietnamese words, odd combinations of vowels and consonants that apply to the bright and eager faces I see spread across a large auditorium. This is the linguistic milieu into which Tami will be thrown if she is successful in pursuit of her chosen career as a college professor. Her struggle with names like *Siphonia tulipa*, *Steganorhynchus dunwoodii*, and *Hoplorhynchus acanthatholius* now seems to have been good training. It's taught her to be patient with odd words whose meanings have significance for you personally. Kinichiro Obata would likely have understood, and greatly appreciated the value to Tami of her etymological lessons delivered at the hands of unnamed parasites living in damselflies.

From the shores of Dunwoody Pond and the gallery filled with ancient words, Tami retreats into darkness to answer her original question about the animals that live inside insects: How do the cell membranes of damselflies interact with those of this community of parasites? She sits before a giant steel machine, her hands on its knurled knobs, her fingers making delicate adjustments. Strange images pass across a screen whose green light reflects off her face. She's at the end of her search, seeing, at last, why she chose damselflies, spent those untold hours at the microscope, cut open so many intestines, struggled with long words and wing veins and markings that showed she'd caught the right species. In the darkness she smiles at the naive questions ringing in her ears, the questions strangers often ask: What good can possibly come out of your work? Satisfied with a picture on the screen, she presses a button and turns the image into a photograph. I chose an insect, and because of it, became a child in wonderland, and will forever be a child in wonderland.

But what *good* comes from your work, they ask again, persistent, unsatisfied. I have produced a child who will forever be in wonderland, she says one more time, firmly, with a touch of new hardness in her voice, losing some of her patience with people who continue to ask questions but don't seem to want to hear the answers. The bandage is gone from the finger that twists the knobs to move her specimen sealed away into a giant vacuum cylinder. This single slice of biological material she's studying represents a year of work and waiting. The bandage represents her lesson in patience with herself. Her parasite is embedded in a block of plastic. She needed to cut that block into a certain shape in order to slice off a section so thin the plastic looks golden. Blocks are trimmed with glass knives. Glass knives are made, broken from squares of quarter-inch plate. Glass knives are sharp as hell; eventually you cut yourself. Then you wait while the finger heals. The diamond knife, used to slice off the golden section, is also sharp, and unforgiving of those who've not learned their first lessons with glass.

The waiting is nothing new. Tami waited until the plastic hardened, and before that, she waited while her specimens took their journeys through caustic chemicals and buffers, each step timed, each step a potential loss of her year's work. Earlier she'd waited for her microscope lessons. On the shores of Dunwoody Pond, she'd waited for summer, for the right species of damselflies, and then for exactly the right species of parasites to appear in the intestines. Then she waited for the parasites to produce spores, and she waited for literature to be sent from across the ocean. She was patient with her own mistakes. Before a specimen could begin its trek from her lab bench to the electron microscope, it had to be fixed in place, attached to the insect gut lining. In order to be of any value, the parasite had to stay attached through all the processing and handling. Many were lost along the way. By the time her parasites were embedded in plastic, they were black, almost unrecognizable versions of themselves. She had no idea whether they would be satisfactory for her purposes. If they were not, then she'd have to wait another year, for another crop of damselflies and another generation of parasites to emerge from Dunwoody Pond.

Her finger healed, she had to wait for "time on the scope." Big expensive scientific instruments are heavily used; she had to schedule her hours far in advance. She spent her waiting hours writing. Once she had an idea, she remembers, then decided to explore it, beginning with some insects she'd caught on the shore of Dunwoody Pond and some questions that came into her mind when she saw the parasites in the insect's intestines. She knew early on that she'd eventually have much writing and waiting to do. If she could return to that first day when she walked into wonderland with her card, *Siphonia tulipa*, and waded through the excited children swirling around her knees, what would she say to herself, and to these kids? Be patient. Science takes time. The distance from Dunwoody to the dark room can be measured in many ways, by the lengthening list of technical skills she's acquired, by the stack of obscure literature accumulating in her files, by the roster of published names of parasites. But the most telling measure of her success at converting herself into a scientist is Tami's patience, her understanding of the strictly human trait—patience—that permeates all work well done.

The lives Tami sees on the electron microscope screen are not so easily analyzed as crushed midge parts in an adult damselfly's intestine. She gets the distinct feeling that what she's seeing first appeared on Earth during the Carboniferous, that the cell membranes so delicately pushed together are evidence of a relationship that's been kept alive through repeated encounters between damselflies and their internal parasites, a sort of reaffirmation of a common need, for two hundred and fifty million years. She focuses on the exact spot where *Steganorhynchus dun-*

woodii and *Ischnura verticalis* come together. The picture looks nothing like she'd envisioned it earlier. Electron micrographs are impressionist drawings, to be deciphered only if you have a clear sense of what you started with; and electron micrographs of parasite membranes up against their host's, when viewed at a magnification of ten thousand times, present one with an additional problem: Which one is the parasite, and which one is the host?

In this case the question is fairly easy to answer. Magnified ten thousand times, all those gregarine parasites with the elegant, geographical, honorific and euphonius names are seen to have row upon row of regular folds at their surfaces. As Tami turns the knobs and moves these folds through the viewing screen, their marching patterns are almost hypnotic. See? Here's where they've been pushed aside by the contact with a damselfly cell; here's where they're cut so cleanly by the diamond knife that you can see tiny tubes supporting each fold; and here's where the spaces between the folds are filled with something dark, making you think of glue, or other sticky secretions. In the electron microscope, a damselfly looks like a beautiful stomachache. Cells lining the intestine are very tall, packed together like the stalks in sheaves of wheat, puffy and fragile, especially at their globular top ends which appear to break open, spilling enzymes into the gut. Tami turns the knobs still further, following the marching parasite folds down into the sheaf of damselfly cells pushed aside by the gregarine's neck. And at the end of the neck, Dunwoody's Lampshadenose's nose: The lampshade is actually a fuzzy balloon filled with mysterious granules. I wonder what those granules are? Tami asks herself in the dark. After all this time, all this waiting, all this work, she's still in the dark, both literally and figuratively, having found only more difficult questions as answers. Once more she smiles. At the end of a long and arduous search and all I find are more difficult questions? Good; I must be on my way to becoming a scientist!

*   *   *

Six years after I handed her a card with the name *Siphonia tulipa* on it, Tami hands me a book, nonfiction, bound in black, its title embossed in gold, and containing 143 pages. This book has four chapters. The events and situations described in those four chapters, as well as the physical appearance of the characters involved, tell a formal story. In places, the language is almost stilted, a requirement of the genre. In other places, the sentences are long, complex descriptions of pictures. The information is not necessarily given in the sequence in which it was acquired; the story has been arranged to guide the reader through the author's series of tasks and thoughts. In the beginning, the main characters are described and we learn something of their history. The second chapter tells about these characters' homes, the environments where they live and die, the disruptions they have to contend with daily. In the third chapter, we get involved in the characters' lives, see how they solve problems in their own unique ways, and find what their common environment forces upon them in the way of compromise. In the final chapter, we see inside these characters, into their innermost secrets, the mysterious and unique traits that keep them apart from other members of their families. Upon closer examination, our impressions from chapter three are shown to be somewhat naive. By the end of this book, a reader knows he's been taken on a journey to places no one else has gone. Tami's name is on the front. My first reaction to receiving it is to place the cover against my nose and breathe in deeply. Brand-new bound thesis copies have a smell that is uniquely intellectual, unquestioningly academic— the smell of major accomplishment.

"Thanks," I say.

"Let's go get a beer," she says.

We adjourn to a local watering hole. A crowd gathers. Tami is on her way to Texas, metamorphosed, like a damselfly, off to bigger ponds, carrying her own versions of *Steganorhynchus dun-*

woodii, *Actinocephalus carrilynnae*, *Hoplorhynchus acanthath-olius*, and *Nubenocephalus nebraskensis*. Somebody asks her if she's going to miss her work, her insects, the High Plains wind, hot muggy mornings around Dunwoody's, and the nighttime storms over the prairie.

"Sort of," she says, then brightens. "But I've got a reminder." We enter a long discussion about the proper sites for your research animal tattoos. Small fish go on the ankle; beetles might be most effective on the back of your hand; frogs are a tossup; snails go on the wrist so that you'll see them every time you check your watch.

"So let's see it," someone asks.

Tami turns so the crowd can inspect the back of her shoulder, where rests *Enallagma civile*, permanently.

3
Nevens Ranch

Where the place?
— T H E   F I R S T   W I T C H

Upon the heath.
— T H E   S E C O N D   W I T C H

Laura came from North Carolina to see if we were good enough, to see for herself whether we matched the recommendation of a friend back home, a good ol' boy who wrote country songs and flew his airplane to Alaska and kept a mind full of stories about unlikely heroes. Go out to Nebraska, he'd said. Then he'd given her a book to read. She read the book, got in her truck with the leaky tire, and headed west. When she got there, she was herded into another truck, with some others, and taken to a place called Nevens Ranch. They passed through the tiny village of Keystone; the scattered houses and outbuildings became more scattered; flower beds, evergreens, traffic, and finally the fences, disappeared until all Laura could see from the truck window were gently rolling hills, low grass with a silver-gray tint, and sky.

The blacktop ends at Keystone, at the "Y" where a turn to the left leads eventually to a heron colony and a turn to the right leads to Nevens, then to another heron colony. East of Keystone the gravel road is deceptively well maintained; the ditches often look freshly graded; burrowing owls watch from the short-cropped pastures; irrigation runoff pools up by a haystack; mosquitoes hide in the hay and drop thousands of eggs into the pools. Even after heavy rains, this part of the road dries quickly, returning to negotiable status within hours. The first warning that maybe the road is not so benign after all comes with the first turn south; the corner is much sharper than expected. Your back tires tend to slip in the gravel again as the next corner, sending you back east, is also sharp, and short. Something's telling you to slow down. Then the ditches start to fill up with poison ivy; blind driveways appear suddenly between large volunteer cottonwoods. Barely marked railroad tracks have a telltale shine on the rails; a freight heavy with coal has hurtled through this way recently; the glint of sunlight on steel tells you not to stop in the middle of these tracks to watch an upland sandpiper on the nearby fence.

Beyond the feebly marked crossing, the road quickly deteriorates, the washboards become brutal and dangerous—at low speeds a driver feels his tie rod ends being jerked from their stud joints; push on the accelerator and tires touch more air than earth. All pretense ends at the Buckhorn Springs ranch gate; past the cattle guard, small ditch willows compress the road into its single lane. From Buckhorn Springs to Nevens, pickups can't pass one another unless someone yields, running right-side wheels up a steep shoulder into the grass. Usually both drivers yield half the road, an unwritten contract whose terms are sometimes carried out rapidly on blind curves. Laura watches this evolution of the road silently, conjuring up all the literature and art she's ever read and seen in order to describe the trail in terms her friends back home might understand. It cannot be done directly and explicitly, she decides. I'm living a metaphor. When we get to wherever we're going, and collect whatever we've come to collect, then we'll go back home the same way we came, over that same road, and things will get better instead of worse.

But at the moment, from her vantage point, wedged into the far back seat, "things" seem to be getting progressively worse in a hurry. The engine whines; she feels the tires lose their grip; tall grass whips the windows; periodically the rear wheels fall into a deep washout, slamming her against the seat. Her companions laugh. Laura doesn't think this trip is very funny. Then she sees the sign—Nevens. Her companions silently watch the weathered board, then resume their animated conversations. What is it about an old railroad sign, she wonders, that makes people stop everything they're doing and read, silently. She twists in the seat, staring back at the wooden "T" disappearing in the distance. Or is it only a sign that proclaims "Nevens" that has such power? If I could capture whatever it is that sign possesses, she thinks, then I'd be a very good teacher.

Nevens Ranch is really the Sillason Ranch. The Sillasons left standing the old sign proclaiming Nevens, a railroad stop before the tracks were moved south down to the floodplain by the river. From west to east, the sign marks a transition in the one and a half lane road, when it changes from brutal to tricky; the pounding suddenly quits, the banks get high enough, and the curves sharp enough, to hide an oncoming pickup, and deeply rutted valleys of loose sand wait in ambush on the uphill slopes. Conversations change, too, past the sign. Theoretical discussions of washboarding effects, sine curves, and the response times of shock absorbers give way to stories about the stupidity of some other driver. He slowed down, they say, laughing, shaking their heads, with a touch of disdain, when he came to the dangerous parts. He got stuck a lot. We had to push him out. You never slow down in a place you might get stuck. Laura listens to this talk and agrees: You never slow down where you might get stuck.

Ten minutes later, Laura finds herself standing in the late morning glare beside a tank, a metal cauldron that boils with minute life. This is it, she thinks; *this* is what we came all the way out here to find? Surely someone knows a secret, she concludes, or an outrageous joke. The secret is that her companions, who have been to Nevens before, now use the word to describe not the Sillason Ranch, nor even the old railroad stop, but in fact a windmill, a watering tank, the overflow pond, and an intellectual state. She looks into the water; green wisps of filamentous scum drift like reflections of the clouds above; hard black beetles stroke quickly through a shadow, then disappear. Above her head, blades turn gently in the soft wind. From a pipe, stuck into the ground, there issues a pulsing stream of cold water that falls into the tank, making a gurgling sound. The gurgling of the windmill pump is the only sound she hears. Then a meadowlark calls, its song an elaborate version of that made by the stream of water. What kind of a place have I come to? Laura asks herself. And where are *Macbeth*'s weird sisters? They should be here, too.

The others from the truck stir the broth, digging, peering into the murk. So Laura waits to see what kind of spectre will emerge from the Nevens tank, what kind of spell will be cast, and who might be afflicted, cursed to become a parasitologist, a pariah of the scientific caste system, someone who not only could identify, but also identify *with*, a blind worm that drank another's blood. The people around her seem to know what they are doing. Crawly things come out of the tank: giant flattened enamel beetles that scratch and dart around inside a bucket, green dragonfly larvae the size of her little finger, that arch their long abdomens and gobble up the smaller wigglers. All this way, Laura thinks, down this miserable road, for *this*? The road to Nevens becomes, in her mind, the measure of her road to the prairies. The blacktop had ended not at Keystone, ten miles back, but somewhere in the vicinity of St. Louis.

But everything Laura sees is at least somewhat familiar to her; these crawly beasts are the currency, the stock-in-trade, of a naturalist. And, like generations of scientists before her, Laura is first and foremost a naturalist. Only the setting, her reasons for being where she is, the people who share, and thus validate, her curiosity, and the meadowlark's call, are new. The habits of one who studies nature are almost universal among the people whom Laura admires, and wants to be with and like. Peering into the Nevens tank, she could easily be young Rachel Carson, hiking the hills and valleys of western Pennsylvania, wondering where the Allegheny River goes, losing all sense of time along the banks of a woodland stream, letting a wet salamander slip through her fingers with a smile. And if her companions today have any influence at all on Laura's future, she'll also be an older Rachel Carson, admitting that "to cope alone and unaided with a subject so vast, so complex, . . . " requires the help of kindred souls. Then, alone at her typewriter, she'll set about to change our perceptions of the world around us.

There is no way to predict who will eventually, and perhaps singlehandedly, alter our collective vision. The futurists tell us that individuals can have an enormous impact on societies, largely, and most permanently, through their writings. Ralph Nader's *Unsafe at Any Speed* rearranged our perceptions of automobile safety and consequently, automobile design; there are no more fins on Cadillacs. Rachel Carson's *Silent Spring* can rightly be claimed to have spawned a nation's environmental conscience; there is no more DDT being spread on our lands, although there is some question whether the replacement chemicals are any kinder to our systems. At least the eagles come in wintertime. And months from now, in the middle of the High Plains winter, Laura will scrape the ice off her truck that still bears North Carolina plates, look up at the sky, and thank Rachel Carson for the bald eagle working its way along the outskirts of town.

But for the moment, Laura is in a field of dunes, standing beside a well tank, somewhere in the middle leg of a long and arduous journey that has no end in sight. The road to Nevens, from North Carolina, is but a step compared to the roads she has yet to travel, and if her experience to date is any indication of what those roads will be—like the one-lane track from Keystone to Paxton with its slippery sands and potential head-on collisions—then her years ahead are filled with a daily struggle to get to some isolated well from which the bug-laden waters flow. She knows that this trail is not laid out upon the ground, nor is it even a symbolic road that wends its way up through the nested hierarchies of corporate America. Instead, her course lies at an ethereal level beyond even the metaphorical. She needs a problem. She needs a problem that she can go to every day, seeking solutions. She needs to find an endless highway she can carry with her and strike out upon each morning.

The solitary nature of this search—for an all-consuming question instead of an all-empowering answer—demands infinite patience from one's associates, impenetrable shielding from the winds that blow inside boardrooms, and a Dunwoody-like toughness in the presence of strangers. But for the moment, at least, she's in the company of like-minded companions turned loose to play at Nevens. While they lose all sense of time, a million miles from television and a thousand years from next semester, I stand a few yards off to the side, watching. My mind wanders as it often does in such circumstances, my thoughts leaping forward into the future to a wonderful evening with dear friends. The home is elegant; fine original art hangs on the walls; a fire burns softly in the fireplace, aromatic wood smell drifts through the room furnished in tasteful earth tones; our host refills my wineglass, sets the bottle aside, settles comfortably in an easy chair next to mine, and asks in all innocence, "What kind of research are you doing now?"

Gulp.

I try to tell him about the microscopic worms that live on the gills of fishes. His eyes start to glaze over. Then I try the story of beautiful one-celled animals that live inside damselflies. Nothing. What about the one-celled animals that live inside beetles? A blank. Worms that live inside frog lungs? By now he's being courteous. He looks around the room. A few glasses are empty. He smiles, excuses himself, retrieves the wine bottle, and makes the rounds of his guests. When he returns, he tells me that my words have not been completely lost. He asks me about my graduate students, what they do, where they come from. Sometimes they come from North Carolina, I say; then I tell him about Laura.

"What kind of a job will she be able to get?"

"She'll be a highly trained person. She'll know statistics. A whole variety of laboratory techniques. She'll be able to write well. She'll be able to teach and teach well. She'll be able to do research."

"All from studying these . . . these parasites."

"Yes." I take a longer sip from the wineglass. "Sure."

"So where will she get a job?"

"I don't know."

"Will she have a teaching certificate?"

"For the public schools? No."

"So what's she going to do with all this training?"

"I don't know."

We reach the end of our conversation about Laura. I change the subject to state agency capital budgets and plans. My host is an architect, a bidder on projects ranging from remodeling to the construction of massive new buildings, and a widely educated man. We talk about current controversies surrounding a state project, the general lack of imagination in local housing developments, and the best churches in town, architecturally speaking. Our drift into church architecture leads to an analysis of the political efforts of the religious right. Inevitably the conversation leads to a review of some fairly stressful local church politics we all know about. That subject reminds me of a comment from a mutual friend who owns a secondhand bookstore. We talk about the mutual friend for a while, about the bookstore, and about the friend's husband's interactions with the person who now holds the husband's former position, now that the husband has retired. We've covered an amazing series of topics, ranging from buildings to economics, from construction contracts to art gallery exhibition policy, from the used book business to theological politics. Surely somewhere in this conversation we have room for Laura and her wild wrigglers. Surely somewhere, sometime, at some intimate dinner party with close friends in beautiful warm surroundings, we can talk about the fact that nobody knows what Laura's going to do, least of all Laura, and how that fact is not causing anyone any misery, least of all Laura. At least for the moment.

Now, as I watch these young people dig through the well tank, completely absorbed in their studies, I think back a few miles to the Nevens sign. Yes. Whatever quality that sign possesses, whatever combination of lines, colors, and patterns gives it the power to capture your attention, is what I need back home. I could use that Nevens sign the next time I get asked about my research, about the mental anguish, emotional highs and lows, thrills and frustrations, experienced by my students. Maybe a piece of jewelry is all one needs, a cast bronze lapel pin shaped like that old railroad sign, with the word NEVENS molded into it, left out in the weather for several years to acquire the requisite patina. You'd know the pin was working if people suddenly stopped talking, stared at your lapel, then for some unexplained reason became more interested in the bark peeling off their fireplace logs than in your analysis of how we could have won the bowl game if only we'd have gotten a few breaks. A pin with that kind of power might be a great gift, a subversive, dangerous gift. But it wouldn't really carry you all the way down the road to Nevens. And when you took it off at night, there's no guarantee you'd be at the end of an allegorical journey to the wild heather in the company of the weird sisters.

Like Dunwoody Pond, the Nevens well site is a teacher's dream, a museum, laboratory, library, and art gallery all combined in one setting and disguised as a windmill, a stock tank, and an overflow pond. What's it really like, people sometimes ask, going out there? It must be wonderful, they say, so much fun! What do you do? Imagine yourself a college student, I reply, majoring in journalism at a school somewhere in the middle of the plains. They smile, thinking "Nebraska." Then imagine, I continue, that instead of the football stadium, it's *The New York Times* that's right next to the journalism building. Their faces usually go blank at that point. No football stadium? Sure, I say, then a journalism major can walk a hundred yards, right into the front door of *The New York Times*, sit down at any desk he or she wants, take any beat from homicide to the fine arts, be editor for a day, copy reader, classified ad taker, cover the Mets or the Yankees, anything. That's what it's like teaching biology at Nevens. Or at Dunwoody's.

Usually people who hear this explanation are still thinking—no football stadium? How can we imagine no football stadium? So I try another tack. Imagine the Supreme Court right next to the law school so that any student could walk right over there, twenty-four hours a day, seven days a week, and play justice, listen to arguments, handle the evidence and make decisions that affect other's lives. The Supreme Court right next to the law school? Sure. Or Congress next door to the political science department. A kid could walk next door, introduce legislation, debate the budget, declare war, impeach a President.

Well, adds Laura, *The New York Times* isn't *right* next door to journalism, nor the Supreme Court *right* next to the law school, nor Congress *right* next to the political science department. Nobody actually *walks* to Congress; you still have to travel the Keystone Road.

And to the young woman from North Carolina, the Keystone Road is as long and tortuous as the highway to New York, or to the Supreme Court, or even to Congress. After having traveled to Nevens seeking her future, her feet rest just inside the edge of a field of dunes. The field is about two hundred miles long and a hundred miles wide. A mile to the south of where she stands, across the floodplain, across a large cattail marsh and hay meadow, runs a river—cold, clear, meandering, swift in places, filled with pretty stones. Five miles south of that river runs another river, this one much more temperamental than the first, less dependable, more capricious, wider, shallower, filled with flowing sand and flashing minnows. Both these rivers come out of the mountains far to the west, then meander across the High Plains for nearly a thousand miles before joining the Missouri. Fifteen miles west of where Laura stands, the closer river has been dammed, making a reservoir, twenty miles long, whose waves wash sand out of the hills and spread it into wide and glaring beaches. A few yards beneath Laura's feet lies a porous sandstone, filled with cold and crystal water—fossil water left behind when the glaciers melted and the mastodonts died. She's on the middle of an underground freshwater ocean. But she didn't travel all this way to study oceans.

As far as she can see to the north, west, and east, stand grasses; regardless of how they're spelled, their names are anagrams for "prairie"—blue grama, sand bluestem, little bluestem, switchgrass, needle-and-thread grass, prairie sandreed. The scientific names, too, spell "prairie" but with a lyrical twist—*Bouteloua gracilis, Andropogon gerardii, Andropogon scoparius, Panicum virgatum, Stipa comata, Calamovilfa longifolia.* Laura knows her plants; she's brought with her from North Carolina the insight to find questions in the grass. How do herbivorous insects influence the growth of grass blades? How, along the spectrum of wet to dry soil moisture, do the various species sort themselves out? Maybe the answers to those questions would somehow explain the patterns she sees on distant dunes, patterns she knows will turn to shades of rust and ochre before she finishes her work in this region of the world. She even understands the underlying genetics of these plants, at least as far as underlying genetics can be understood by a well-educated person holding a college degree from one of the country's prestigious universities. Some have two sets of chromosomes, some have three, some have four, and some have many; but in the case of grass, it's not always clear how many species are involved in all of these chromosome number variants and some claim that we're seeing instant evolution at work among the plants that blanket as much as a fourth of the Earth's surface. Instant evolution is interesting, but the good ol' boy who wrote country songs and flew his plane to Alaska didn't send her out to the prairies to find problems she'd already found.

Through the course of the summer, we come back to Nevens, like we come back to Dunwoody, because these places are reliable teachers. You can design a whole day's work around the animals you know you'll find there. Who participates in the lifetime events of a worm that lives inside a frog's lung? What kind of snails, what kinds of insects, do frogs eat? We think we know the answers to those two questions, but in confirming the answers, twenty people spend twenty hours at Nevens first, then late into the night, in the laboratory with snails, worms, literature, and computers, finishing the business all scientists eventually have to finish, namely determining whether you've asked the right question. The student from North Carolina studies her classmates at work, listens to their talk, their answers, their questions, their interpretations of the observations they've made, and analyzes the effect that Nevens is having on them. She goes back along that road, watching the ritual at the well tank over and over again, then returning also along the same road, studying the sand traps and blind curves.

After a month of examining the well and its overflow pond, and watching her fellow students get hypnotized each time, she relegates Nevens Ranch to the Domain of Sacred Places. This is not a site where you begin a serious research project, she decides. No, Nevens is a country where you go to get away, to reestablish in your mind the fact that you once decided to become a biological scientist, and to take others who need to be dragged out along the Keystone Road for their own good. Spend enough miles on that road and Nevens starts to look more like your reward for passing tests in other habitats than like a place on which to hang your future. Dunwoody's may be your doctoral comprehensive exams, but Nevens is your museum. Nevens Ranch is the place you buy when you have enough money to buy anything in the world. Standing at the Nevens well tank, on the hill above the floodplain, you can see signs of water in the distance—the cattails, a treeline along a spring on the adjoining ranch, and sometimes sun reflecting off the North Platte stream itself. The proximity of all this shallow water in the midst of dune country has biological significance: Animals that live *in* water occur throughout the Sandhills. The Nevens well tank is not quite the biological equivalent of a southeastern farm pond, but it's close enough so that the young woman from North Carolina, looking for new horizons, looks beyond the windmill and the weathered gateposts. She didn't travel the Keystone Road to find something familiar to her.

"I always took the best teachers and the hardest courses, so I could get a good education," she says during one of our conversations.

Good, I think. What you need is a problem to match your past experiences, a problem that at once serves as both the best teacher and the hardest course of any you've encountered in a young life spent looking for the best and most difficult. I have a mental file drawer full of such problems; they get handed out only to the most fearless of students. None of these problems can be "solved." Instead, they must be carved into shape, like the finest marble, wrestled into form, like a long novel, or woven into a pattern, thread by hard-bought thread, like a Navajo rug.

"The dragonflies are very diverse here, and predictable. Now if you really want a problem, you could work on the one-celled animals that live in dragonflies."

"Tami and Aris are already doing the damselflies."

"The dragonflies are an order of magnitude more difficult than the damselflies." As a minimum, you have to catch dragonflies. Laura's watched twenty people spend two hours at this task and come up with only a few. Neither adult nor the aquatic immature dragonflies are as heavily infected as the damselflies. You never find dragonfly naiads in the large numbers that you find damselflies. All of the odds are stacked against anyone who wants to study the gregarine parasites of dragonflies. Elucidating the life history of any one species of these parasites would be a problem worthy of someone who always took the hardest courses and the best teachers. Besides, their parasites are in even more of a taxonomic mess than those of the damselflies.

"It's too seasonal." She's quickly evaluated the task and discerned the one aspect of it that cannot be resolved by brains and effort.

"You can always come back here in the fall. I mean, it's a public road, right? Dragonflies ought to be out until October. And the Sillasons have given us permission to use the place."

She stares south, across the miles, toward the hills between the North and South Platte Rivers. Anyone can see she's filed the Nevens well tank away in a folder entitled Domain of Sacred Places.

"You can even come back here in the winter if you want

to. If the naiads are here in the summer, they're bound to be here in the winter, under the ice. I'm sure the Sillasons wouldn't live here if they couldn't live here. They have to get to the grocery store. How can winter be that bad?"

She gives me that look, the look of a North Carolina person who knows full well that a Nebraska Sandhills winter can be "that bad" and then some.

"We have a bunch of theoretical predictions that need to be tested."

She perks up.

"Take the one about the schedule at which infective stages enter the system. That's really a doctoral problem, but if you chose the right animals, it could be done at the masters level." If you don't choose the right animals, theoretical problems can't be done at any level. I sense she knows that already. "Or the one about the diversity of parasites in populations of hosts. That one's magnificent, but again you need the right animals."

"What are the right animals?"

"Beetles, probably." I give it some more thought. "Yeah. Beetles."

She perks down.

"We have a rule of thumb back in North Carolina," says Laura, "that you never study anything smaller than your thumb."

Everything we study lives in animals that are smaller than her thumb. When we do get our hands on larger animals, they're most likely road kills. Laura's come fifteen hundred miles to discover that all our rules of thumb violate her rule of thumb.

"We don't do birds anymore."

She nods and agrees. Federal regulations surrounding the study of parasites in wild birds make meaningful research very difficult. Add the logistical demands of bird research, and such work becomes nearly impossible. Besides, I learned a long time ago that homogeneity is the key to success when one studies the parasites of wild animals. Birds are highly heterogeneous; compared to some insects, birds' lives are extraordinarily diverse; birds move long distances through a great diversity of habitats; most of them are opportunistic feeders; they live a long time; by the time you collect enough birds for statistical analysis, they've lived too long, sampled too many habitats, been too many places, to be called homogeneous in any sense of the word except taxonomic.

"We've done a lot of preliminary work on the rodents, but I've never been confident we could study them without destroying so many that you'd mess up the populations." By "study," of course, I mean catch, dissect, and convert into worms, fleas, lice, ticks, mites, intestinal and blood dwelling protozoa. The Nebraska Sandhills are packed with small rodents, a beautiful array of species.

Laura loves small mammals. In North Carolina, flying squirrels were the animals larger than her thumb.

"The only problem with rodents," I continue, "is that the conceptual basis for what you'd be doing is really better studied using some other kinds of animals."

"For example?"

"Beetles."

"Back in North Carolina, we have this rule of thumb," she reminds me.

"I know."

"The caddis flies are pretty impressive, though." She pauses. "Speaking of insects." As part of her course requirements, Laura has started a small project on the ciliated protozoa that live on the surfaces of caddis fly larvae. The larvae build elaborate tubular houses out of sticks or stones, depending on the species. I'd always wondered whether the filter feeding protozoa were confined to certain parts of the larva's body, a distribution that might be related to the water currents passing through the house. The caddis flies Laura's chosen live in clear spring-fed streams flowing out of the Sandhills and into the North Platte

River. Like everyone who comes to places like Nevens and Dunwoody's, she had total freedom to select her class project. Given total freedom, she chose to violate her own rule of thumb.

"Well, give some thought to the river," I reply. Thesis problems are to class projects as rivers are to spring-fed streams. "The River" doesn't need to be identified. We both know "The River" means the farther one of the two south of where she stands—the wider, shallower, more temperamental and capricious river, the one filled with moving sand and flashing minnows and problems fit for someone who always took the hardest courses and best teachers. Such problems seem to flow out of the mountains in an inexhaustible stream. Sometimes I wonder what kind of a dam it would take to shut off the supply of biological questions that ride through our country on the South Platte's ripples. She stares for a long time across the floodplain.

"They're both just mountains washing away to the sea," I continue.

In the near distance, the North Platte River's icy water rolls stones down from the Medicine Bow and Laramie Ranges, cowboy movie landscapes that are one of America's most persistent geological anomalies. The stones are granite, basalt, quartzite and slate slivers, and tuff all mixed, ground, and spread into a river bed; when irrigation season is over, and the North Platte is once again safely trapped behind a series of dams, rock hunters walk the bed looking for the best products of the Laramide Orogeny, or the Laramide Revolution as it's sometimes called. Nevens Ranch, Inc. property runs up to the edge of the North Platte, although "edge" is probably not the correct word to describe the tangle of young cottonwoods and willows, temporary algae-laden pools, prickly pear flats, cattails, and shifting gravel bars that sort of mark both north and south boundaries of the river. In the late summer, Sillason machinery and cowboys hay the floodplain. The mowers generally stop when they get to the river. There are no official markers at the edge. It's not easy to get to the North

Platte River without walking a great distance across places like
the Nevens Ranch.

Over the horizon, beyond the North Platte, the South Platte
carries Colorado away to the Gulf of Mexico. Packed in the South
Platte's sand are stone points and ceremonial knives, deer bones,
patent medicine bottles, bison teeth and rusted out early sev-
enties car bodies. The South Platte is the more accessible of the
two rivers. Numerous private roads cross the railroad tracks that
enter the floodplain from Nebraska Highway 30. All the towns
along I-80 are connected to the interstate by long bridges over
the South Platte, and almost every bridge has turnoffs near each
end where you can park. Beneath these bridges you find graffiti,
burned out tin cans, occasional items of clothing, and piles of
stained toilet paper. Making your way down to the water, you
step through and around this kind of material, wondering who
the nomads were, and whether the abandoned encampments
you'd have encountered two hundred years ago would have reminded you of the present ones. Probably not, you decide; ar-
rowheads and bison teeth have a dignity that disposable lighters
and used condoms will never acquire. But the South Platte is
filled with minnows—smart, capable, streetwise, social, abundant, i.e., thoroughly modern minnows. And like minnows everywhere, these have parasites.

"We'll look at some minnows tomorrow if you want to." You
can look at about anything you want to tomorrow, if you're a
young scientist searching for your future in the Sandhills. "We've
never looked at the minnows very carefully. You know, in any
kind of organized way."

"Look at some minnows" means hours at the microscope
for Laura, in just the same way as "look at some———" has
meant hours at the microscope for them all.

"Sure," she says, "I'd like to look at some minnows." "Minnows" and hours at the microscope mean yet another violation
of her rule of thumb.

In her search for the best teachers and hardest courses, she's enrolled in ichthyology as well as parasitology. On the days she's not cutting open animals looking for other animals, she's in the water, sometimes up to her chin, seining every river and stream within a hundred miles of Camelot. Minnows have acquired a certain stature on Laura's rather uncompromising list of big challenges. They are small, like she is, and sensitive, intelligent, subtle, and curious. It's not easy to get to know a minnow. It's even more difficult to get to know the worms that live in and on minnows. The idea takes hold. The river is her next stop. Nevens, a place that seemed at first so distant, odd, isolated, now becomes a memory of her first serious encounter with the deep prairie and the microscopic worlds that occupy its seeps.

By late March, the river ice is mostly gone and in a dry year the South Platte usually is safe to seine. So we pile in a van and head west. I turn off the interstate at the Paxton exit, cross the South Platte River, assessing its safety, and drive slowly through town. All the familiar landmarks are intact—Ole's Big Game Bar, the grain elevator, and the ball diamond. Paxton seems fixed in time; even the state patrol car, parked in the same place for a decade, hasn't moved. North of Paxton we climb a long hill. From the top you can look out over the North Platte River valley; beyond the treeline lie the Sandhills, and just inside the sea of dunes, Nevens.

In late March, the Nevens well site shows all the marks of a "not that bad" winter—bleak, beat down red-buff grasses, dried sunflowers, seedheads pushed into the dirt by hooves, low water, and a windmill whose blades are locked parallel to the gales. Laura'd looked at some minnows the previous fall and found her problem. Two closely related species of the genus *Notropis* live close enough together in the South Platte River to be caught in the same seine haul. Yet they have very different parasite faunas. Laura asks: Why? Observation plus question equals problem. This equation has held true since Aristotle walked the streets of Athens, at the height of Classical Greek culture, twenty-three hundred years ago.

We're actually on our way to a research site seven miles west of Paxton along the South Platte River, but have chosen a twenty-mile detour past the Nevens well tank. We arrive within hours of spring. Frogs lie belly up in the shallows, or dark and seemingly dead in the hoofprints stomped into the mud around the overflow pond. In your hand they warm up and start squirming. We work our way around the pond with nets, sampling for snails and dragonfly larvae. With every step, Laura's decision to keep Nevens in the Domain of Sacred Places is validated; Nevens is a place to "do biology" but not necessarily a place to do research. Then she spots a movement in the weeds. A quick sweep of the net and she's caught a very large tiger salamander. Stuck to its abdomen are two exceedingly healthy leeches. Laura's found a friend—Appalachia is the heart of salamander country, and North Carolina's in the heart of Appalachia. She surveys the surrounding prairies. Her better judgment and training take precedence over her first thoughts: There must be a biologically legitimate avenue for tiger salamanders to get to the Nevens well pond; this symbol of my home country couldn't have come here the same way I did—down the Keystone Road. She lets the animal slip back into the water. But if it did get to Nevens down the Keystone Road, then surely it's earned its spot on the High Plains.

4
Options and Constraints

Though boys throw stones at frogs in
sport, the frogs do not die in sport,
but in earnest.

— PLUTARCH

I don't see no p'ints about that frog
that's any better'n any other frog.

— A FELLER

Marv sits and ponders the fates of toads:
Should I cut up one, or let it live long enough so that the worms
in its lungs will start laying eggs, which in turn will get coughed
up assuming toads cough, swallowed assuming toads swallow
their own coughed up phlegm, and passed in toad feces? He
decides to spare his toad, test his assumptions about worms laying eggs and toad coughing, spitting and swallowing, and dig
through toad manure rather than bloody lung tissues. For if Marv
is sure of one thing in life, it's the fact that toads produce manure;

he's been cleaning up after them for months. Furthermore, it's November now, and the value of live toads escalates in proportion to the dropping daily temperature.

And if by chance he finds a worm egg, then that toad will indeed have a "p'int about" it that "makes it better'n any other" toad. Such a "p'int" will be immediately understood by Marv's friends and fellow laborers, who sit calmly, chewing on their sandwiches or digging microwaved casserole out of plastic containers, listening intently to his talk about toad worms, lungs, blood, coughed up swallowed phlegm, and feces, nodding agreement, offering suggestions, asking questions, and in general acting vitally interested in Marv's various options and constraints while oblivious of the rather unappetizing vocabulary he's using to describe his work of love.

Marv's present dilemma is a product of a chance observation that stuck in my mind and gnawed away—a parasitic idea—until it metamorphosed into a full-blown problem. We'd been at Nevens all afternoon, catching dragonflies, and had spent the evening on the subtleties of population biology. As we were cleaning up the mess, a student asked me a question: What are these things in the lip of this dragonfly naiad? The naiad was one we'd caught in the well tank; it was about two inches long, green, active—a member of the genus *Anax*. Adults of this genus are large, swift-flying, insects; they make me think of Carboniferous jungles, imagining salamanderlike amphibians six feet long slithering away through the rotten organic muck that will become our present-day coal and natural gas supplies. The Carboniferous is gone, but organisms persist, especially the dragonflies that evolved then. Dragonflies saw the giant amphibians disappear, to be replaced by modern frogs, toads, and salamanders; dragonflies saw the dinosaurs evolve, flourish, then become extinct, leaving us with the birds they became; dragonflies saw the sabre-toothed cats evolve, along with rhinos and woolly mammoths that multiplied, spread out over the prairies, and become extinct. Dragonflies of the genus *Anax* saw the glaciers come then retreat, watched the giant bison herds thunder across the plains with Indians in hot pursuit, and buzzed the rain puddles not yet filled with weeds that formed when earth movers dug the footing trenches for an interstate highway bridge. The student who showed me the naiad's lip thought she'd made a discovery. For her, she had.

A dragonfly naiad's lower lip is a hinged, extensible, prey-capturing structure that's usually kept folded up under the head. At the end of this lip are some teeth. The student brought me the slide with the lip neatly dissected, laid out so she could see the identification characters. Through the microscope we could see these characters plainly. But the lip also was filled with parasitic worms, embedded in the tissues, and pulsing slowly within their cysts.

"Those look like frog lung flukes," I said. Frog lung flukes were supposed to be in the dragonfly naiad's rectum, not in its lip.

"Can we do an experimental infection?" she responded.

I thought about the question, so simple and naive yet so insightful, and filled with hidden potential. Amphibians have probably been getting accidently infected with lung flukes for at least a hundred million years and now a student wants to do it once more, on purpose.

"Sure," I said. "The only problem is you can't find a frog that you know is uninfected."

Experimental infections require a supply of uninfected hosts, and in Western Nebraska, almost all frogs have flukes in their lungs.

"We could try a toad," she said.

I always listen to the naive; they are the source of some of our best ideas. We had looked inside local toad lungs for fifteen years and never found a lung fluke. Toads and frogs differ in many ways, and in our part of the world, the contents of their lungs is included in that difference. Twenty-four hours later, she had a handful of small toads and a bucket of infected dragonfly naiads. Toad in one hand, dragonfly in the other, she proceeded to reenact an event that has occurred uncountable times, over vast regions of the continents: Amphibian eats dragonfly.

"When should I cut them?" she wondered.

"How strong is your curiosity?" I asked.

"Pretty strong."

"Two hours," was my advice. If a worm can't get out of its own cyst inside a toad stomach in two hours, then something's wrong. Two hours later, she showed me the toad's esophagus. At least two hundred tiny immature worms just emerged from their cysts were charging upward, anatomically speaking, toward the lungs. Three days later, she looked inside another toad. The lungs were packed with worms, still immature but rapidly growing. When it came to excysting and migrating to the lungs, these worms could not distinguish between frogs and toads.

At this point, a whole range of possible problems entered my mind—problems that would be nice opportunities for future students. In order to tackle one of these problems, all you had to do was learn how to keep tadpoles, toads, frogs, several species of snails, dragonfly naiads, and a whole assortment of other aquatic invertebrates, alive in large numbers at the same time. Once you'd learned how to do that, then you could begin to think about studying the worms. So when Marv arrived on the scene, I mentioned the experience with frog worms living in toads and with worms in an insect's lip instead of its rectum. These parasites' options were obviously greater than their constraints.

"That's interesting," he said. "What species were they?"

Ah, the overriding question in biology: What species is it? We hadn't finished the experiment; summer ended before the worms grew up and we didn't have time to infect more toads. You can't tell what species the worms are unless they're adults.

"I don't know," I admitted.

"Now that *would* be interesting," he continued, "if some species could infect toads and others couldn't." Like all good young zoologists, Marv was very familiar with the frog lung-fluke life cycle, at least as that cycle was presented in the textbooks. Larvae in a dragonfly lip instead of its rectum and frog worms in a toad, however, seemed to be violations of the constraints implied by the published information. But he also knew there were several species of frog lung flukes, and not all of them had been studied extensively. Marv decided immediately to repeat the experiment that seemed to reveal parasite options not previously understood.

"Give it six months," I suggested. If you're going to spend the rest of your life studying parasites, then the first six months spent testing your ability to run a zoo is rather well-invested time. The toad whose feces are the object of such intense study is one product of that investment. This toad is one of three, the other two of which have begun to pass worm eggs, a sure indication of successful infection. The third one, however, raises questions in Marv's mind. And if there is one thing Marv could use at this point in his life, it's questions. He's trying to write fifty of them as a means of organizing his thoughts.

Conventional wisdom says scientists seek answers to questions. But science founders on conventional wisdom because the most creative scientists tend to chase unconventional wisdom, and when they catch that wisdom, it usually takes the form of questions rather than answers. Robbers would be easier to catch, thinks Marv, than questions. When two of his toads started to pass eggs, the blessed event stimulated much excitement among some of the most intelligent and well educated people of our society. But he still doesn't know exactly which of several related species of worm is producing the eggs, thus the discussion of one particular toad's fate. Marv's now ready to go chasing unconventional wisdom; the questions, however, are his major vehicle and they lie, still undiscovered, somewhere in the back of his subconscious.

Having chosen to commit his time and talents to a struggle with problems manifested in the domain of toads, frogs, and worms, Marv joins a long tradition of human intellectual involvement with amphibians. From middle school students getting their first look at the insides of a vertebrate animal, to the elegant, classical, and sometimes controversial, studies carried out on amphibian embryos by the turn-of-the-century and prewar German scientific elite, students have turned to the lowly frog for laboratory materials. Untold numbers of surgeons made their first cuts as thirteen-year-old children in biology class, on a frog. Uncounted legions of family physicians did their first experiments with living tissue in a college physiology course, using a frog's leg muscle and an electrode.

But just as Marv is beginning his tortuous journey through the lives of worms that live in frogs, he finds himself wondering if maybe he will be the last person ever to do so. In his morning paper, he reads a story about the mysterious and precipitous global decline in frog populations. The story contains some speculation; a few scientists proclaim that the lowly frog might be humanity's equivalent of the coal miner's canary. In the mornings, Marv struggles with the implications of evolutionary history. Were there worms in frog lungs during the Mesozoic? Probably so, he concludes. In the afternoons, Marv struggles with the possibility that he's chosen to work on species made rare by human activity. And at noon, he props his feet up on his desk, takes a sandwich out of a sack, and ponders the meaning of a life caught between the primeval and the catastrophic—the ancient lush wet marshes and the leaking ozone layer above his head.

Although there's a possibility Marv will be the last person to write a doctoral thesis on the worms in frog lungs, he's certainly not the first individual whose mind became entangled in knotty and arcane problems surrounding the biology of amphibians. Early in this century, Paul Kammerer, an Austrian biologist, conducted a series of convoluted experiments that have never been repeated. The subject of Kammerer's studies was the midwife toad, *Alytes obstetricans*. Midwife toads mate on land. The male grabs the female around the middle and hangs on until she releases the eggs, whereupon he fertilizes them. Then he gathers up the eggs he's just fertilized, winds them up around his hind legs, and carries them around until they hatch. Thus the origin of the specific epithet: *obstetricans*. Many male frogs and toads develop so-called nuptial pads on their front palms, spines or ridges that function to help the male maintain his hold on the female. In the teleological phraseology of Western culture, the male toad "needs" these pads to keep his hold on the slippery female in the water. *Alytes obstetricans*, however, which mates on land, does not "need" pads, and so does not develop them. Kammerer reasoned, however, that if the midwife toads mated in water, the males would develop the nuptial pads they "needed." "Need" is a forbidden word in modern evolutionary thought. The implied purposefulness, linearity, and logic that permeates the Kammerer story are also scorned. But in pre-war Europe, among the biologists who studied amphibian evolution, the real object of scorn was Paul Kammerer.

Kammerer claimed that when he made midwife toads mate in water, the males indeed developed the nuptial pads. Then Kammerer committed the ultimate blasphemy: He also claimed that such nuptial pads, once developed, were present in the offspring of toads that mated in the water; i.e., the newly acquired pads were inherited. In the 1920s, in Europe, to claim inheritance of acquired characteristics was to side with the historical pariah of evolutionary biology, Jean Baptiste Pierre Antoine de Monet, aka Chevalier de Lamarck. British scientists in particular, buoyed by their proximity to Darwin, lunged to the attack. Eventually Paul Kammerer, artist, poet, highly sensitive handler and lover of frogs, toads, and salamanders, publisher of lengthy, detailed, and beautifully illustrated papers in the leading German scientific journals, a man with the infinite patience to suffer through years of trial, error, frustration, and labor for a few successfully breeding pairs, succumbed to the British assault on his integrity. To quote his sympathetic biographer, Arthur Koestler:

> In the early afternoon of September 23, 1926, a roadworker found the dead body of a well-dressed man in a dark suit on an Austrian mountain path. He was in a sitting position, back propped against a vertical rock face, right hand clutching the pistol with which he had shot himself through the head.

The note in Kammerer's pocket suggested his body be sent to a university where it might be used in dissection. He hoped such service might offset some of the complete worthlessness that, according to his critics, characterized his scientific work. But the critics, especially the English evolutionist William Bateson and G. A. Boulenger, Curator of Reptiles at the British Museum, never displayed the sensitivity and patience of Paul Kammerer, at least with respect to the breeding of midwife toads. The biological question of whether midwife toads develop inheritable characteristics when forced to breed in water remains unresolved. Although the world did not stop turning when Paul Kammerer pulled the trigger, his case remains a model worthy of study, especially if one embarks on an intellectual journey involving the husbandry of amphibians. One wonders just how many scientific questions remain beyond our reach, their answers requiring too much sensitivity and patience on our part. Marv with his noontime sandwich and daily homage to frog and toad manure, studies the Kammerer case carefully. In a world where reputation is built on accomplishments others can repeat, he's not interested in being the only one on Earth who can manipulate the lives of frogs, toads, worms, snails, and various insects, all at the same time, in order to answer one of the fifty questions he has yet to write.

But Marv doesn't have to look to prewar Austria to find antecedents in the amphibian parasite business; the desk upon which his feet are propped was once occupied by another young man, named Lee, who launched his own career with a study of tapeworms that live inside *Bufo woodhousei*, the Rocky Mountain toad. Lee once spent three field seasons stalking the floodplain of the South Platte River weekly, beginning in May and ending when the cold north wind told him there would be no more toads until spring. His descent into mystery, however, would begin in mid-June when he had to wait until nearly ten o'clock for the sun to go down, and the nights were warm, sometimes with thundershowers. Then he would gather up his headlamp, check the battery, and head for the river where the toads were coming out of the sand.

Only after the three years had passed, and Lee was about to move on, having written his thesis and been admitted to a prestigious medical school, did he admit that walking the river alone, night after night, living the life of a Rocky Mountain toad, listening to the noises, the steady gurgle of the river, midges swirling around his headlamp, did unusual thoughts come into his mind. You think about things you'd never otherwise think, he said. Like what? I wondered. There are forces at work that nobody really understands until they've walked the river at night for three years, he continued. It's stupid to admit this, but you feel like you don't really belong there. You know? The river seems suspicious of you, and everything else that's hidden out there in the night gets quiet as you walk by, and then suddenly, there's this toad in your spotlight. You feel like you're in another time, like maybe millions of years ago. I reminded him of the trucks on the interstate a mile away. You learn to not hear the trucks, he said, out on the river looking for toads at two or three in the morning. And the trains? A major two-track line runs by the barbed wire gate to the property where Lee did his night work. You can think of trains as dinosaurs, he replied, literally as well as metaphorically.

So I went to the river with him periodically. I concluded you had to do it every week, alone, in order to get the sense of not belonging. But once Lee needed to do an experiment, to determine approximately how many prey items a toad less than an inch long needed to eat in order to get one tapeworm. In order to do this experiment, he had to find about a hundred noninfected one-inch toads. The experiment was relatively simple in design: dissect a third of the toads to make sure they were not infected, set another third aside until the end of the experiment, at which time they would be dissected to make sure the toad supply stayed uninfected, then take a third of them to the river, put them in a large enclosure, and let them get tapeworms on their own. Lee searched the countryside until he found an uninfected population of toads, collected his hundred, and started his experiment, whereupon he also became obligated to feed shoeboxes full of newly metamorphosed toads. He accomplished the feeding by collecting midges—delicate mosquitolike flies that don't bite— from the window screens, sucking them up into a tube, then emptying the tube into the shoeboxes.

One day as I watched him at this activity, I commented on Lee's animals.

"They look like they're waiting for you, like they're lined up against this end waiting for you to bring them their midges," I said.

"They are," he replied.

A typical Rocky Mountain toad egg mass contains twenty thousand eggs. There are thousands of egg-laying toads along the river, and millions of eggs. With the June emergence, you have to walk carefully along the South Platte to avoid stepping on newly metamorphosed toads, many with tail stubs. And how many midges or other insects does a three-quarter-inch-long toad eat? As a result of his experiment, Lee concluded that on the average, each newly metamorphosed toad eats at least four thousand ants, rove beetles, ground beetles, spiders, and midges a week. They also acquire new tapeworms at the rate of about one a week, so in general, one insect in four thousand is infected with a larval tapeworm. During the month after they emerge from the tadpole stage onto dry land, the overwhelming majority of these toads get tapeworms, at least those along the river. By late August there are also millions, if not billions, of tapeworms. A reasonably sized fraction of the river's biological productivity ends up as tapeworms in toad intestines.

At the end of his three years, I reflected on Lee's nighttime strolls along the South Platte River.

"Yes," he admits, "it was spooky at first."

"I guess you got used to it." After all, he did get his thesis written and published.

"I got used to it, but it takes a while to get over the feeling that you don't belong." He grinned. "Then it gets even more spooky."

"More spooky?"

"Yes. When you feel you belong on the river at midnight and not back here with the rest of the humans."

I mentioned the Kammerer book. He'd already read it.

There are no paintings of frogs or toads on the walls of Altamira, but certainly by the time of Plutarch in the years immediately following the Crucifixion, humanity's less noble interactions with the Anura were well understood. Plutarch's observation of boys and frogs carries a hint of sadness, a prophetical assertion about the relationship between maturity and our view of nature. Operating on a more proletarian scale than Plutarch, however, Mark

Twain may have come closer to understanding the bond between animals with certain properties and humans who seek questions to answer, i.e., between student Marv and a supposedly infected toad who, for the time being, remains nameless, but nevertheless has a p'int about him that makes him better'n any other.

The p'int, of course, is that this toad's been fed a small crustacean with an infective stage of a lung worm in it. The important p'int concerns this small crustacean as much as the toad, however, because the worm was not supposed to infect the crustacean. Furthermore, the worms that Marv used to infect the crustacean came from a snail he wasn't sure he'd be able to infect. And to top it all off, the toad is not exactly the right animal for these worms to be in, having been chosen more for convenience than for biology because toads seem to be easier to keep happy in the lab than frogs. Marv's deliberately doing everything he can to violate his expectations, and doing it quite successfully; so at the moment, he's sitting in the middle of more options than constraints. And for this reason, his list of fifty questions is getting more difficult to write daily.

His task is not made any easier by the history of our knowledge of worms in frog lungs, a history that for nearly two hundred years after the worm's discovery was filled not only with ignorance about how worms get into lungs, but also a perplexing lack of interest in the problem. The great Dutch zoologist Jan Swammerdam was evidently the first to record an observation of frog lung flukes. Swammerdam at the height of his career could easily have been mistaken for a graduate student like Marv—dissecting the lives of bees and mayflies, dissecting insects with scissors so small they had to be sharpened using a microscope, dropping the altruistic, and presumably lucrative, study of medicine for the more personally satisfying pursuit of anatomy and in the process ruining his health.

But in the three centuries that have passed since Jan Swammerdam cut open his frog's lung, Marv's options for reporting his observations have become relatively constrained. Some of Swammerdam's most important research was not published until fifty years after his death. But if Marv is to earn enough of a scientific reputation to find employment three years hence, he must first find the right questions to answer. All scientific reputation rests on the right questions. Good answers to trivial questions don't propel an individual very far, nor very fast, along a productive scholarly track. And if this pursuit of reputation sounds at odds with the idealism, the love of nature, the elegant distance naturalists are supposed to maintain from the workaday battles, then listen to Marv's wife Liz, sitting beside him at a local watering hole on a Friday afternoon.

"The problem isn't that you guys aren't going to get a job," she says with that certain clean analytical lilt to her voice. "You can always get a job. That's not the problem. The problem is that you might have to get a job that won't let you do what you want to do."

And around the table there are solemn nods, an acceptance of the difference between earning a living and earning a life of the mind, as well as awareness of the machinations and hoop-jumping that so often accompany what Marv has come to view as the most ideal of lives: that of the college professor. All the choices that he's made during the past five years—what to study, how to study it, where to go to study it, and in what arena to display the results of his endeavors, have steadily narrowed his paths and directed the career that he has yet to live. He's followed his interests, which have led him to problems involving parasitic worms. Like most problems with worms, this one has a long, convoluted history involving some notorious characters with now famous names. And, in an event Marv hopes to reenact within a modern conceptual framework, that history also involves a young scientist, Wendell Krull, who was perplexed not only by the worms themselves, but also, perhaps even more so, by the fact that "the life history of more of these forms has not been the subject of inquiry." That is, Krull was mystified as much by the fact that so few people had asked interesting questions as by the convoluted lives of the parasites themselves.

Thanks to Wendell Krull, we now have a general idea about the way frogs are supposed to get worms in their lungs, namely from eating dragonflies that are themselves infected with larval worms. We also have a good idea about how the dragonflies get their worms, namely from sucking them up into their rectums. There are about 5,500 known species of dragonflies and their close relatives, including about half this number that are delicate species called damselflies. Virtually all of these insects spend their immature lives in water. Dragonflies are also described as hemimetabolic, i.e., they undergo gradual changes as an immature stage with every molt; in the last molt they acquire their wings. The aquatic stage, called a naiad, is a highly predatory animal with a toothed lower lip that serves as a flashing extensible jawlike prey capturing device.

In an aquarium, dragonfly naiads can decimate a population of smaller insects and crustacea, consuming even small members of their own species. The only condition to being eaten is that the prey item passes through the naiad's field of vision. Dragonfly naiads will snap at anything, including forceps, tips of sticks, and needles. But all this unmerciful viciousness at the anterior end is matched by what some would consider a touch of surreality at the posterior end: rectal gills. Dragonfly naiads breathe by sucking water into their rectum, which in turn is highly folded and functions much like fish gills. The rectal water-sucking and expulsion muscles are powerful; naiads can shoot through the water like a torpedo by forcing the water out of their rectum. But it's the sucking that brings in the parasites.

The worm eggs that get coughed up by the frog, swallowed, then passed out with frog feces, must be eaten by a snail before the parasite can enter the next stage of its existence. Inside the snail, a lip pops off one end of the egg, aided, no doubt, by snail digestive enzymes eating away at the seal that holds the lid in place. Within the snail intestine, a tiny ciliated larva, called a miracidium, emerges from this egg, and immediately begins burrowing through the snail's intestinal wall. Once it pierces the intestine, the miracidium then burrows into the snail's liver, where it sheds its ciliated skin and begins producing more worm larvae.

This next stage of the life cycle is called a sporocyst, and its offspring grow within its body, originating in balls of cells. Eventually these balls of cells will themselves become larvae, in turn producing, within themselves, still more balls of cells in an asexual embryonic malignancy of reproduction that can consume much of the snail's body and leave it sterile. The life of a frog lung fluke, like the lives of the other six thousand species of parasitic flatworms called trematodes, reminds one of the old nursery rhyme riddle about going to St. Ives. Except in this case, the worms aren't going to St. Ives, but coming the other way, and like the cats in the sacks of the seven wives, increasing exponentially in numbers.

St. Ives is a town north of London where, in medieval times, merchants gathered in an annual fair along the Kings Highway to hawk iron, goat skins, boards, oil, and ale. In the nursery rhyme, the traveler to St. Ives obviously meets someone who's found plenty to acquire at the fair, although having left with seven wives, each with seven sacks full of cats, you have to wonder whether the poor soul enjoyed too much of the famous ale while he was there. The snail upon meeting the worm egg finds itself in about the same situation as the traveler might have had he invited the man with seven wives—each with seven sacks, each sack with seven cats, etc.—not only to turn around and join him, but also to share his table. That is, the snail picks up a load that soon begins increasing, making even the seven sacks, each with seven kits, seem light by comparison. Eventually, however, the snail will be relieved, at least symbolically.

Eventually. After a month or two, the sporocysts, actually the daughters of the original larva in the egg, begin producing another stage of immature worm—a cercaria—but also within themselves. So instead of containing a liver full of sporocysts containing sporocysts, the snail has a liver full of daughter sporocysts containing cercariae. The cercariae are active, tailed, little creatures that burrow back out through the snail, breaking free into the water.

Scientists have been mightily fascinated with cercariae for a long time. These flatworm larval stages are among the most frenetically active of animals, spinning, jerking, vibrating at high speeds, and when the violent lashing around stops temporarily, kneading themselves incessantly, especially their anterior ends, using powerful muscles. But it's not the larval motion that has captured the attention of biologists; instead, generations of parasitologists have sought evidence for a correspondence of sorts— a structural correspondence—between the cercariae and their adult worms that still lie one or two life-cycle stages away into the future.

There are many practical reasons, of course, for a scientist to wonder whether he or she can identify a parasitic worm from its larval stages. Such knowledge serves to help determine whether human populations are at risk of being infected with flatworms, especially blood flukes of the genus *Schistosoma*. Humans become infected when cercariae burrow into the skin. Some experts estimate that schistosomaiasis—the disease condition resulting from infection—affects one out of every ten people on Earth, mostly in the tropics and subtropics. Chronic blood fluke infection clogs the liver, destroys the bladder, and can turn even a young child into a withered, bloated shadow of a human. If you can collect snails, and identify the cercariae that emerge from them, you can also, in theory, determine the human risks and, if necessary, design disease control programs.

But beyond the obvious epidemiological importance of

some larval worms, and of the knowledge necessary to identify them, the cercarial stage evidently has a charm all its own, an ability to fascinate people and a capacity to demand the attention of a scientist for the duration of his career. And the source of this power is fairly obvious: A cercaria is like a crystal ball; if you can identify it, you can see into both the past and the future. Capture a snail, isolate it in a small jar of water, find the cercariae that emerge from that snail during the next few hours, identify the larvae, and you know what animal has passed by the pond and dropped feces the snail fed upon, and you know approximately when that fecal drop occurred. You also know what the worm will become when it metamorphoses into an adult, although in order to make such an identification you must study structures that are not present in the adult. Identify the cercaria correctly and you know the other species it must encounter to survive, the schedule of its reproduction, the environmental conditions that allow its own descendents to reproduce in turn.

The cercariae from one of Marv's parasites seem to have many options and few constraints—they burrow into almost any crustacean or aquatic insect, and thus become infective for frogs that eat such prey. Cercariae from his other parasite species, however, are much more constrained; so far he's been unable to get them to infect any of the insects and crustacea he's able to maintain in his lab. The parasites' species are supposed to be closely related. They have undoubtedly occupied amphibians, sometimes the same individual lung, for at least a hundred and probably two hundred million years. Yet their eggs do not develop in the same species, in fact not even in the same families, of snails. And their larval cercaria do not encyst in the same families, or even orders, of aquatic insects and crustaceans.

By late spring following his November with toads, Marv's managed to domesticate three species of parasites, but through his observations of who infects whom, he's suddenly witnessed the shadow of an event that may have occurred a hundred million

years before he was born—namely the divergence of two evolutionary lines. In pursuing the naive experiments done years
earlier by a naive student asking honest questions, Marv's managed to wrap his hands around an enormously difficult, complex,
and mysterious problem in evolution. Now, in order to keep this
problem in the lab, he maintains a menagerie. The first stages
of his three species of parasites must be passed in a snail. But
which snail? Marv has nine different kinds, all breeding in the
lab, and all mothered along with pride.

"*Gyraulus parvus.*" He taps the side of an aquarium bearing the sign "*Physa* Kills." When species of *Physa* are allowed
to get a foothold in the *Gyraulus* tank, the latter disappear. *Gyraulus parvus* is the tiniest and most delicate of Marv's snails.
You can see their hearts beating through their translucent
flatly coiled shells. A *Gyraulus* whose shell is a quarter inch in
diameter is a giant. His stocks of *Gyraulus* came from Nevens
Ranch. *Physa* and *Gyraulus* live together in nature; but Marv
thinks that in the aquarium *Physa* eats the *Gyraulus* eggs.
"There must be more places to hide at Nevens than in the
lab," he concludes.

Promenetus exacuous is the other delicate, tiny, planorbid;
Marv's supply of *P. exacuous* came from the Buckhorn Springs
Ranch, "next door" to Nevens—a mile away, that is. *Promenetus
exacuous* is the most beautiful of all his snails—translucent, like
Gyraulus, but with a sharp edge to the spiral. "They've laid all
these little egg masses." He points them out; I take his word;
without a hand lens you can't see the eggs.

"*Physella gyrina.*" *P. gyrina* also came from Buckhorn
Springs Ranch. I remember the day we collected these two—*P.
exacuous* and *P. gyrina*. We were so excited about finding *Promenetus* that we left our best aquatic net leaning against the side
of my station wagon. Then we drove off. Two hours later, down
on the river, we missed the net. So we drove back to Buckhorn
Springs along the Keystone to Paxton Road, Laura's road to Nev-

ens. When we got to Buckhorn Springs, the graders had come through and driven over the net.

"I wonder what that driver thought," says Marv. " 'Hey! There's an aquatic dip net lying in the middle of the worst, most isolated road on Earth! I bet some parasitologists have been here!' "

"I'm guessing he didn't see it," I venture.

"Not see an aquatic dip net six feet long in the middle of the road?"

We leave the conversation at that.

"*Helisoma anceps*." *Helisoma anceps*' aquarium also bears the "*Physa* Kills" sign. *Helisoma* is a large, robust, flatly coiled snail many times the size of *Gyraulus*. I have no idea why it can't live with *Physa* in an aquarium. Marv's original *Helisoma anceps* came from Dunwoody Pond. Piece by piece, species by species, he's trying to reassemble the Sandhills inside a city building.

"*Planorbella armigera*." Marv's especially proud of his *Planorbella*; they're flat, pretty, superficially similar to *Helisoma*, but with their smallest coils raised slightly on one side. His colony of *P. armigera* came from a place called Atkinson Lake.

"*Physa virgata anatina*." This one breeds in gallon jars; the stock came from the Popelka farm in Pawnee County. "*Physa virgata virgata*." Another gallon jar; *P. v. virgata* came from the South Platte River near Roscoe. Marv has become our resident snail expert. He had to order a doctoral thesis from Michigan to obtain the information necessary to distinguish between these two subspecies. Then he had to learn to dissect out the male reproductive system in order to use that information. The snails I saw him working on were less than a half-inch long.

"*Stagnicola elodes*." His aquarium full of *S. elodes* came from an oxbow on the North Platte River north of Paxton; these snails could be descendants of the ones I wrote about in 1976, viewing them as migrating pioneers looking for their times and places.

"*Fossaria cockerelli.*" *F. cockerelli* is a brand-new one for me, a snail species I'd not seen in nearly twenty years of looking for snails. Laura collected some in a mudhole along the South Platte River near Ogallala, then took me to the exact spot. I'd been in that mudhole many times but had never seen *Fossaria cockerelli.* But I'd not been in the mudhole this year. They must have washed down out of the Rockies, I conclude. Marv shatters my excuse.

"Look at this," he says. He's holding a 1906 volume of the Philadelphia Academy of Natural Sciences Proceedings entitled *Mollusca fauna of the American Southwest.* "Some guy named Simpson found *Fossaria cockerelli* in the South Platte River at Ogallala a hundred years ago." Marv grins. "He probably got it from the same place you did."

That comment drives me crazy. So, I spend a day in the library trying to bring a dead man to life. Charles Torrey Simpson was born in 1846 and died in his home at Lemon City, Florida, on December 17, 1932. In between he managed to do what we do: collect snails from the river at Ogallala and measure them. When Charles Simpson visited my collecting site, Ogallala's main claim to fame was as a rail head for Texas cattle drives. Simpson was associated with the Smithsonian Institution. Nowhere in all his writings does he mention going to Nebraska, with one possible exception—an 1890 paper entitled simply "Shell Collecting," published in a journal named *The Nautilus.* Eventually I give up trying to find the man who waded in my river a century before me and left behind some proof of his interests. Maybe he didn't wade there after all; maybe some cowboy sent him a few specimens with a note written by the local schoolmarm: What is this? None of his friends felt it necessary to compose an obituary, and the volumes of *The Nautilus* from 1890 are missing from our library. One more time, I think, when you lose library resources and museum collections, you lose your ancestors—

intellectual and otherwise—and along with them, your sense of who you are.

At the other end of the lung worms' life cycles are their amphibian hosts. The spadefoot toads came from a roadside ditch on Nevens Ranch. The spadefoots are still tadpoles; adults come out of the ground to breed only when the thunderstorms are bad enough. So far, June's been a good month for bad storms. The tiger salamanders came from Nevens well tank, along with one of the leopard frog species, *Rana pipiens*. Marv's other species of leopard frog, *R. blairi*, came from his grandfather's farm, along with the tree frogs, *Hyla versicolor*. Of all his animals, I appreciate the tree frogs most; their favorite music is Mother Mabel Carter's bluegrass version of "When the Band is Playing Dixie." When Mother Mabel is on the lab stereo, the tree frogs start screaming. *Rana catesbeiana*, the bullfrog, was named for Mark Catesby, a Revolutionary War–era naturalist who came to America to shoot birds, paint pictures, and find a rich wife; and succeeded in doing all three. Marv's bullfrogs started out as tadpoles. All these adult amphibians are carnivorous. Marv feeds them crickets dusted with vitamins; he hatches about two thousand food crickets a week, from colonies maintained in specially constructed boxes. Some crickets escape and find their ways throughout the building. Now you can hear crickets calling from the molecular biology labs. Singing crickets are prettier than the dead cockroaches in the hall and they sing no matter what music is on the stereo.

By the end of his first year, Marv has written his fifty questions, brought much of the life cycle into his lab, captured three species of frog lung parasites, and wrapped his hands around an enormous problem in evolutionary biology. But there is one constraint he's unable to avoid: At the post-snail stage of its life, one of his parasite species is evidently highly specific to dragonflies and does not have the option of entering crustacea or other insects. Regardless of all his skill and efforts to bring nature into the laboratory where he can perform experiments, Marv is still at the mercy of an evolutionary constraint deeply embedded in the genetic makeup of a parasitic worm. Dragonflies are among the wildest of insects; like all "top predators" they require a massive and interdependent food pyramid beneath them. And like all of the wildest beasts, they require large amounts of space and time—space to dart and course in search of prey, time to develop through their long aquatic immature stages. Evolutionary "choices" made during the Carboniferous haunt a young scientist at the end of the twentieth century. He knows that if he can work around this problem, the rest of his career will be relatively easy.

"I need to go back to Nevens," he concludes.

We get in the car and drive west. For six hours we talk about crustaceans, dragonflies, frogs, snails, evolution, academic politics, national politics, international politics, the local newspapers, teaching, good books, religion, cloud patterns, low-level radioactive waste, ranching, center pivot irrigation, surface water management, computers, statistics, and art. I find myself wishing I could capture the effortless transitions this conversation seems to take. If I could do that, then it would be easy to teach science the way science ought to be taught, integrating it into the whole of human experience. As a teacher I have to keep hoping that the problem of not being able to make these transitions in class, is mine and not the students'. Yet Marv has done something that most of the others do not, or cannot, or will not. Because of his willingness to take a gigantic gamble, he's entered a world that is rich, almost beyond description, with metaphor, history, literature, and scientific struggle, just by choosing to "work with frog lung flukes" and repeat an experiment done by a naive student who had to leave for home at the end of the summer and never got to look inside another toad.

Part II

The Biogenetic Law

We struck up a conversation at a party. You know I was in the first class you ever taught, she said; it was at seven-thirty in the morning, on Tuesday, Thursday, and Saturday. She was waiting for a telephone call while we talked; a death was imminent. There must have been three hundred people in that class, she continued; you were awful young; what I remember most is that phrase "ontogeny recapitulates phylogeny." Three hundred and sixty-two, I replied, remembering the assignment as clearly as she did; and yes, everyone remembers "ontogeny recapitulates phylogeny"; it's poetry that seems to stick in your mind. The phone rang a couple of times, and each time she cut off the talk and quickly turned her head toward the kitchen. We're expecting it to happen tonight, she explained; he was terminal before we left to come out here. When the call from her son finally came, she was gone for several minutes, then returned resigned to an event over which she had no control. We talked to people all over the country, she said; no one could help. What's he going to do with the body? I asked, trying to be helpful. He's going to bury it, she answered. You could put it in the refrigerator, I suggested; I'd be happy to sit down with him and do an autopsy tomorrow. He's too emotionally involved, she said; he'll just bury it. But no more lizards. The vet I talked to in Florida said it was a parasite and that nobody around here knew how to take care of lizards. Get him a microscope, and a jar of weeds and water, I said. That's what my dad did to me when I was ten years old. He's got a microscope, she said. He's got everything he needs to be a scientist when he grows up.

5
Ontogeny

Making variations on a theme is the
crux of creativity.

— DOUGLAS HOFSTADTER

Where do they come from, these young men and women who go into the field to search for questions to answer? Who are they? What did their mothers and fathers say to them as children? It's all right to like butterflies. It's okay to read in bed. Sure, I brought you all that scrap paper from the office so you could draw pictures on it. See those stars; do you know how far away they are? They're so far away that you have to think in a completely different way to even imagine such a distance. You can't think of miles. You have to think of years, and of the distance light can travel in a whole year, if you want to imagine how far away is a star.

What's a bedtime rule in comparison to the stars, huh? You can't leave the kitchen light on if you want to watch the night sky from the backyard. That's right, but when you go back to turn off the light, then there'll be all these moths on the screen. A kid at school said that if you put a big moth with thin antennae inside the window screen, then another one, even bigger, with big fuzzy antennae, will be there on the outside, in the morning. Maybe so, but you have to catch the first one and you can't hurt it. So you have to decide whether to watch the stars or try to catch moths. No, you don't have to decide between those two things. You just have to stay up late enough to do both. But then the stars move. They move every night, so you can stay up late two nights. Maybe every night. But sometimes it rains. And lightnings. You know something? Lightning is just electricity; the ancient legends claim thunder is the sound of giants bowling up above the clouds, but it's really only the sound of lightning, a giant electric spark.

Wait a minute. An electric spark? Lightning is only an electric spark? Thunder is not people bowling in the clouds, but just the sound of a giant spark? That means you could just go outside in the rain and listen to the thunder and you'd be hearing through a microscope. You know how sparks have a quick little spitting sound? You take a battery, just a flashlight battery and a wire, and you turn off the light. Then hold that wire against one end of the battery and scrape the other end of the wire against the other end of the battery, and if you look real close, and your eyes are used to the dark, you can see this tiny spark. If you listen, and there's no television set on and no dog is barking, then you can hear a tiny crackle. But lightning is just what that spark would look like if you were so small that your whole city could sit on the end of the battery. And thunder is what that crackle would sound like if you lived in that city. So all you have to do is watch the lightning and listen to the thunder, and instead of your getting small, sparks and crackle have gotten big, so you can see them magnified, and hear them magnified. And suddenly you appreciate the meaning of the word energy—that fundamental construct of the universe—simply by watching nature and talking about it, and acting like you're a scientist, a curious, searching, *scientist*.

Yes, where do they come from, these *scientists*? They come out of this mass of flesh we call humanity. They come walking in, asking questions just like they must have done as kids, except that by the time they show up at my door, their questions have grown up, too. But something else has happened to them before they surface. Or maybe that "something else" is what caused them to surface in the first place. If I had to guess, after all these decades of experience, I'd say that "something else" had to do with their relationships with authority. The people who come into my office seeking difficult questions to ask have lost most of their respect for any authority that would try to control their minds. And once they enter, whatever remnant of such respect that might have lingered in their souls quickly dissipates in the pungent air. They get sent into the lab. Soon Dante's ringing command—"All hope abandon, ye who enter here!"—appears, in computer-generated Vivaldi font, taped over their door. Every day they walk under that small sign and smile. The hope they've abandoned is the hope of being undertested.

This quality of seeking complex puzzles as life-organizing entities is in stark contrast to the guiding forces that influence so many others. At the opposite end of the rebellion spectrum from my graduate students stands a massive body of humanity, exemplified beautifully by five young women and one young man who also passed through my life recently. Although I know very little about them, they have stimulated me finally to tell someone about them, and this essay is an ideal place to do just that. The women range in physical appearance from gorgeous to somewhat plain; the young man is black. They probably come from a variety of backgrounds. I'll assume they are representative of our general population, but a 1/6 black/white ratio is a little higher than you'd expect from a random sample of Nebraskans. However, these children are no random sample; they share a trait that makes them stand out from the rest of us: They are brilliant. They also share a second trait, but that is the frustration in their story. Five of them have decided to become physical therapists, the sixth, an occupational therapist. But instead of being physical therapists, which I'll allow that an aging and almost psychotically athletic nation needs in fairly large numbers, any of them could be a university president.

The institution of higher learning where I hold an endowed chair gives away scholarships, called Regents Scholarships, that pay a student's tuition for four years. Unlike an athletic scholarship, which cannot be revoked because of injury or failure to make the starting lineup, the Regents Scholarships can be revoked if a student's grade average falls below a 3.5 (on a 1–4 scale, with *A* being equal to 4). At least three of the five young women mentioned above hold four-year Regents Scholarships. On the surface, they are among our best and brightest. We have selected them for their promise, and invested tens of thousands of dollars in their education. What will American society get in return for its investment in these young citizens if we help them achieve their stated goals? The answer is: four docile, perfectly behaved, nonthreatening females who spend their days "working with" the torn anterior cruciate ligaments of our high school football heroes, one minority member of our society who does the same, and a final, highly articulate and intelligent woman, who does "occupational therapy."

The young black man, it seemed to me, represented a double tragedy. He registered for my introductory biology class, enrollment 265, along with three or four other minority students. In such classes, at exam time, I generally give two forms of a test, coded somehow, so that students sitting next to one another do not have exactly the same exams—the oldest, and perhaps most easily subverted, trick of the teachering trade. Hour exams are multiple choice, computer graded. When I give more than one form of a test, I go through the answer sheets before delivering them to the grader, to ensure that none get graded by the wrong key. On the first exam in this particular class, there were several students who had not written their code numbers on their answer sheets, so I looked at their answers, and tried to make a decision based on my memory of the correct sequence of a, b, c, d, or e. As it turned out, this one minority student was among those who had not written his code number on his answer sheet, or so I thought.

Some teachers return 265 answer sheets after a test by dumping them in a box outside their offices and letting students paw through them at will. I tend to hand back tests personally, usually by visiting labs. I do this not because I'm a good, or especially sensitive teacher, but because I'm an angry one. I want to find out who my better students are. Two years hence they will come to my office asking for a letter of recommendation for medical school and they seem to resist my desire to learn their names, preferring in most cases to remain anonymous, especially if they are women. But I have this idealistic feeling that teachers should, in obeyance to some abstract law of the profession, put a face with a name. And when the faces are black, at least at this university, the names that go with them are very easy to remember. As I handed back this young man's test I noticed, in my own handwriting, the word "high" on his answer sheet. I had decided which key his test would be graded by. As he took the paper, I also noticed his grade: the lowest in the class, so low, in fact, as to be far below statistical probability had he guessed at every answer.

The combination of his low grade and my own handwriting on his answer sheet continued to nag at my mind over the next few days, so one morning I waited in the back of the lecture hall and stopped him as he entered the room. I asked him to bring his answer sheet by my office so I could check it by the other key. In a few days, he arrived. Sure enough, the test had been graded by the wrong key, and when I looked closely I realized he had written the code number on his answer sheet, as instructed, but in a place that made it easily overlooked. When I corrected the mistake, his score went from the lowest to one of the highest. While he was in my office, we had a conversation that went something like this.

JJ: Where'd you go to high school?

Student: Prep. I like those philosophical questions like you brought up in class today. We used to talk about those kinds of things in high school. I like that kind of stuff.

JJ: What's your major?

Student: Physical therapy.

By "Prep" he meant Creighton Prep, one of Omaha's premier schools, although all male, and run by the Catholic church. Prep has outstanding athletic teams and sends many high academic performance students to the university. We talked a few more minutes. I asked what other courses he was taking, how his grades were, and so forth. When he left, I thought: Here's a very articulate, intelligent, minority student with a great deal of presence. A born leader. He needs to be in a position of power.

In introductory biology courses, the second exam, on genetics, is usually a killer. The mixture of math and biology seems to baffle many people. Statistically, a full third of the freshmen in a typical class have great trouble carrying out the following calculations in a biological context:

$$\frac{1}{2} \times \frac{1}{2} = ? \ (A: \frac{1}{4})$$
$$\frac{1}{4} \times \frac{1}{2} = ? \ (A: \frac{1}{8})$$
$$\frac{1}{4} \times \frac{1}{4} = ? \ (A: \frac{1}{16})$$

When genetics is the subject, my class average typically drops from about 70 percent to about 60 percent. Although his attendance had fallen off between the first and second test, the minority student from Prep did very well on the genetics exam.

He always sat by himself, next to a wall. Between the second and third tests, his attendance became poorer still; I hardly ever saw him. And he may not have attended at all between the third test and final exam; I simply don't remember. But his grades stayed high, and when he came to the exams, he sat alone. I watched him carefully, as I do all students whose grades are somehow inconsistent with their attendance. He was perfectly honest. He would have made a $B+$ in biology, a grade that a hundred white students would have killed for. I say "would have," because when I filled out the computer-generated grade roster sent by the records office, I realized he had registered *P/NP*, i.e., pass or no pass.

Pass/No Pass is a grading option allowed on a limited number of courses. One must make at least a C equivalent to get a P; the P is not figured into one's overall grade average. Pass/No Pass is intended for people who want to try out a course they think they might be interested in, but because of background, are afraid of a low grade. No department allows the *P/NP* option for major courses. Long before I encountered this minority student under conditions forced on us both by economics—the "classroom" with 265 students—someone had convinced him to register *P/NP* and choose physical therapy as a career goal. I have worried about that single student every day since I saw that *P/NP* option on my grade roster and filled in P beside his name, worried that I did not handle the situation correctly. Should I have called him into my office again, told him he had a responsibility to the rest of humanity to set his goals much higher than physical therapy? Is it my decision to make, to tell such a child— *someone else's child*—that we all need for him eventually to become president of our university instead of its athletic trainer?

That is a good question. That is a very good question.

Why did I begin this exploration of the origin and development of America's scientific human resources with an obscure case study on grading, advising, and career choice among college freshman? Because in this instance, in a nation crying desperately for brains, ideas, leadership, and women and minorities in positions of major responsibility, my smartest women and the most promising black student I'd had in class in more than a decade had opted to become physical and occupational therapists. And the minority student had registered *P/NP* for a course in which he'd coasted to an honest *B+* with a miserable attendance record. None of these students had looked to complex puzzles for intellectual stimulation and a life focus; instead, all had sought ways to satisfy authority—in this case me—and to become certified for an immediate job. Who, I wondered, had been giving these young people advice? Parents? Ministers? Like-minded friends? Relatives? High school coach/counselors? All of these students had set their sights well below their abilities. Who had told them to do so? And why had the students listened to such advisers?

I never saw the minority student again. Insofar as I know, he's disappeared into oblivion. The women, at least, had been obedient. I'd asked them to stop by the office, and when they did, I quizzed them about their plans, i.e., major, career choices, etc. Out of the five, three got angry when I told them they'd set their sights too low. One of the other two laughed. Then she came back to ask for a letter of recommendation for a major scholarship, which she received. The fifth listened politely as I ranted and raved about a nation that sends its brains to be physical therapists and its demagogues to Washington. Then she said: You're right. I never saw her again, either.

Although in at least one of these cases attendance may have slipped, none of the six students ever actually walked out of class. But in the large auditorium where I lecture to nearly three hundred young people, someone always walks out when I start talking about evolution. A few always walk out when I start to show slides, too, but I've not decided whether it's for the same reason. No, I think they walk out on my slides because they don't believe I will, or can, ask questions about them. The only time they don't walk out on slides is when the pictures come from their books, i.e., as part of the promotional package publishers give to faculty members who adopt their textbooks. I interpret this difference in behavior as respect for the text, or more properly, as respect for the *authority* of the text. Thus when I tell my students, "This slide is Fig. 14–2 from your textbook," I sense that they immediately feel they might be tested on it. I also pick out five- or ten-minute video clips from various tapes, usually ones made using our microscope camera, but also sometimes using my own camcorder. They never walk out on video, no matter how amateurish. Instead, they settle down in their chairs, smile and pay rapt attention to the many monitors hanging from the ceiling. I've not tried a home video on evolution.

Only one time has a student actually told me why he walked out when I started talking about evolution; it was late in the semester when he came to me with a form to sign in order to drop the course without receiving an *F*.

"I was raised in a religious family," he said. "God created the Earth in seven days. The Earth is only a few thousand years old. That's what I was taught and that's what I believe." I'd not even asked him why he was dropping, but just assumed from his grades that something was not going right.

His words are as direct a quote as any other in this book. They sound almost stereotypical, almost contrived for the purpose of making a point in writing, but they're not. I didn't ask the kid why he'd come to college; our professional codes of conduct demand that we smile and shrug, and wish them well.

In addition to such conversations, university professors receive student feedback from a variety of sources, one of them being evaluation forms. Virtually all college and university courses in the United States are evaluated by the students enrolled, typically through use of a relatively standardized form. At our university, the faculty member is not supposed to be in the room when these evaluations are filled out, nor is the faculty member supposed to handle the completed forms before they are delivered to the department secretary. I am, by implication of policy, considered too intimidating and untrustworthy to be present in the room or in possession of the unprocessed forms. So on one day near the end of the semester, I hand out the forms, ask a student to collect and deliver them to the department secretary, and leave the auditorium. Three or four months later, I get the forms back from the administration, along with a statistical summary of the responses.

I usually study the summaries before deciding the circumstances under which I'll read the written commentaries on the backs of the pages. After good score semesters, I read them immediately. After average or below average semesters, I go home, pour a big glass of some stiff beverage, settle into a chair in the privacy of my own living room, and make some pretense at being analytical about their impressions of whatever I've brought into their world. Particularly memorable comments are either saved or remembered. "Make Janovy a janitor" is one such remark. As director of the Cedar Point Biological Station summer field program for seven years, I heard that phrase repeatedly—from my wife. An important administrative position in which I was responsible for the feeding, housing, safety, financial records, and academic programs of a mini-university left me quite often with a broom in hand late at night. After a year or two of this smart aleck but affectionate and understanding joke derived from a student evaluation form, my wife realized I was on the handle of a broom so often that the phrase had lost its punch. Then she started saying it when I ended up with a mop, instead of a broom, in my hands. We both understood the words to be a subtle commentary on the illusion of power and the reality of responsibility that accompany administrative positions.

But the most bewildering words I've ever read on the back of a student evaluation form were the following ones, evidently provided as explanation of why someone had walked out when I mentioned evolution: *I don't believe in the Cambrian explosion.* The person who had filled out this form had summarized fifteen weeks of my serious efforts to convey the wonder, importance, facts and concepts of science in general, and biology in particular, with those seven words: *I don't believe in the Cambrian explosion.* On the front of the form are the series of questions to which students respond with a mark-sense dot beside a number. This individual had given me the lowest possible scores on every item. Is, was, it really true that just because he or she did not believe in the Cambrian explosion, that lack of belief translated into the lowest possible evaluations of me as a teacher? Was there a connection? But more importantly, for the question of where young scientists come from, *is* there such a connection between absolute and literal religious belief impressed upon children from an early age, and one's ability to acquire and process information about the only planet in the universe *known* to be inhabited?

The Cambrian explosion is an observed fact, as well as a powerful metaphor and a major part of the conceptual framework of biology. The observation is that of an apparently rapid increase in the diversity and complexity of animal life during a relatively short period of time, geologically speaking. This explosion of life forms occurred about 600 million years ago. There are many explanations for this observation, and not all of them are evolutionary. That is, the seeming suddenness may be a geological artifact resulting from the destruction of pre-Cambrian rocks. Alternatively, the absence of hard parts in early, but complex and diverse, animals may have prevented fossilization and denied us a broad window on late pre-Cambrian times. Regardless of the source of the "explosion," there is ample evidence that half a billion years ago, animal life was wonderfully diverse and complex, and that groups still alive and abundant today, e.g., arthropods and echinoderms, were quite plentiful and present in taxonomic profusion then. The Cambrian explosion is walking-around knowledge for the young scientists whose stories are told in these pages.

The Cambrian explosion also should be walking-around knowledge for most well-educated citizens, especially those who've spent serious recreational time in a natural history museum. Knowledge of the geological record is as much a part of the modern human's perception of the planet that supports us, and the universe that contains that planet, as are the phases of the moon, thunderstorms, oceans, and the Milky Way. An understanding of evolution, and a sense of all that the word implies, are not evil traits, nor should they be considered nonspiritual ones. Instead, such understanding increases one's appreciation for the uniqueness of our planet, the uniqueness of its life forms and habitats, and indeed the uniqueness of the human species. Like the physicists' confrontations with subatomic particles, the biologists' study of evolutionary history elevates and inspires our wonder at the workings of the universe, regardless of where it came from. For some biologists, just as it is for some physicists, such education strengthens and confirms a belief in a Supreme Being while at the same time emphasizes the symbolic and allegorical qualities of the scriptures. Although the study of science sometimes pushes an individual toward atheism, there are many other intellectual pursuits that are capable of similarly influencing one's attitude toward received knowledge. Being a beat reporter on a major newspaper is one of them.

But even for those whose spirituality is strengthened by understanding, their cynicism is also increased by encounters with students who still, in the mid-1990s, deny the existence of a fossil record. Such denial is not a product of education; instead, it's a product of authority, *human* authority. The young man who dropped my course because he believed the world was made in seven days back around 4004 B.C., like the people who walk out at my use of the word "evolution," has been harmed irreparably not by his spirituality, but by those humans who've used spirituality to exert power over a young person's intellectual development. There are plenty of scientists with fairly strong religious beliefs; religion *per se* does not produce docile and obedient college students who choose careers far below their mental abilities.

I don't worry so much about young people who walk out on my slides; I've fallen asleep during slide shows myself, including one given by my own father. But I never fell asleep in a museum, and I stayed wide awake when people handed me valuable fossils to touch with my own hands. The Cambrian explosion came alive when my father ordered a set of prints from the Hughes Tool Company; the pictures were reproductions of paintings depicting the various geological periods and they brought to life the trilobites he'd shown me in some roadcut rocks. I remember his quiet pleasure when he carefully opened the large envelope and pulled out the sheets, one by one, explaining them, staring at them, smiling. I still have those prints and the envelope they came in. The date on the postmark is April 22, 1947; at the time the postage to send that packet of wonder from Houston to Oklahoma City was fifteen cents.

The students whose stories are told in the other chapters of this book have been shown those prints, too, and reacted with great appreciation for their intent. Our present knowledge of the Cambrian period makes the paintings seem rather dated, but that fact also illustrates the basic nature of science. Our observations change continuously, both in quality and in quantity, and the conceptual tools used to interpret those observations also change constantly. Scientists try to do the best with the information and instruments at their disposal, and the instruments include theoretical ones. Sometimes we fail to recognize the importance of our own work, but very often that failure is a product of our own ideas, our own preconceived view of the universe, not of the observations themselves.

That was certainly the case when the Hughes Tool Company artist looked to the fossil record for inspiration. In 1947 the Burgess Shale fossils lay in museum drawers, mislabeled, misidentified, and misinterpreted because of Charles Doolittle Walcott's insistence that they belonged to known taxonomic groups. Walcott was secretary of the Smithsonian Institution and one of the leading and most influential scientists of his day. His story, told in Stephen J. Gould's *Wonderful Life*, is a compelling one, revealing not only Walcott's "mistake," but also the enormous pressures he endured as a leading presidential adviser. But Walcott *did* collect and preserve his materials, depositing them in an institution where they would be available to future generations of legitimate scholars. He knew his rocks were valuable and he knew why they were valuable. His scientific training and tradition of investigation led him to place ultimate "authority" in nonhuman, neutral materials of the universe—pieces of 600-million-year-old Burgess Shale—instead of in his own words, regardless of how strongly he believed them. Had this not been the case, he likely would have tossed the fossils and declared his monographs to be the "facts." Instead, science has "tossed" Walcott's conclusions and interpretations—a fate all scientists expect for their work—and declared his collections to be the material of lasting value.

This story of Charles Walcott and the Burgess Shale has certain similarities to that of Walter Rothschild and his Tring Museum, although the two individuals were strikingly different personalities, and the institutions they served had quite different roles in their respective societies. Instead of being director of a long-established national museum, as Walcott was, Rothschild built his museum, beginning at the age of seven, with family money. In the beginning his collections were the indulgence of a family whose financial power seemed to have no limits. Walter traveled the world and sent hired collectors to those far corners he couldn't reach. The Tring Museum grew in size and importance as its numbers of specimens, particularly of butterflies and birds, increased far beyond the limits first imagined by the Rothschild clan.

But in the end, it was the study and interpretation of these specimens by another generation of scholars, not Walter Rothschild's science, that helped shape our perceptions of the universe. Museums contain evidence of what the world is really like, regardless of what humans say it is like. All science, not just classical zoology and paleontology, collects evidence of what the world is really like regardless of our beliefs and assertions. And routinely, such evidence is examined later, in light of new discoveries. The stuff—the numbers, measurements, tissues, rocks, water, slides, bottles, pinned insects, and mammal skins—is the authority; the humans derive their authority from whatever they do with their stuff.

The trail of this scientific material from its collection to its most current use is sometimes a convoluted one. The literature on evolution contains pictures of now extinct native Hawaiian birds used to illustrate the correspondence between bill structure and feeding habits. In the past, when I used this example, I'd tell my students the story of Walter Rothschild whose collections provide irrefutable evidence of what the world was like in the early years of this century. Today's young people live in a different universe, biologically as well as politically, from their great-grandfathers. But an image of their great-grandfathers' natural world is preserved in current textbooks due to Walter Rothschild's determination to collect original observations, as well as his sale of Tring bird collections to the American Museum of Natural History. The financial proceeds from this sale allowed Walter to pay off two blackmailing mistresses. The intellectual proceeds have allowed generations of scholars to *know*, to *know for certain*, what the world was like before they were born. Knowing what the natural world was like before you were born is knowledge essential to your assessment of the impact your activities are having on the ecological systems that support you. In order to pay off his paramours, Walter Rothschild sold that knowledge. That information ends up in textbooks because the American Museum of Natural History bought it.

No student has ever said *I don't believe in the Hawaiian honey creepers*. But I don't see much difference between not believing in the Cambrian explosion and not believing in the existence of the honey creepers, or of denying the impact human populations have on the environment that supports humans. Such acts of disbelief are at the heart of decisions to walk out of a class because of something a teacher says. And, to a somewhat lesser but nevertheless important extent, a similar disbelief is behind the decisions made by the five young women and the minority student mentioned above who chose career paths leading far below their intellectual capabilities. There is ample evidence that bright kids grow up to become leaders in many societies, not just ours. But such children have to see that evidence in an objective and rational way, and come to understand that the evidence applies to them as well as to the people written about in books. Acts of disbelief originate when humans force their own explanations on the universe without ever trying to answer a complex question about that same universe. To tell a brilliant young minority student to "go major in prephysical therapy so you can get a job, but register for biology Pass/No Pass because it's a hard course" is an act of disbelief. Such acts deprive us of our most valuable resources—our brains, knowledge, understanding, and rationality.

Any one of the six students could have become a scientist easily; none did because someone had told them, quite convincingly, to be physical and occupational therapists instead and they never questioned the instructions. In essence, they were saying: I don't believe in the power of the mind I was born with. They did not accept as valid the raw evidence of their own grades, just as other students have denied the raw evidence of a fossil record, even when handed a specimen. All were told, either explicitly or implicitly, by some human authority, to *do* something and to continue to *do* that something for the rest of their lives, rather than ask the kinds of questions young, including very young, scientists ask. Those questions are phrased, for example, as, *What is the Cambrian explosion, really?* instead of as, *What should I believe about the Cambrian explosion?* And when the answer is given, the young scientist asks, in return, *What is the evidence and where can I find it?* The others either ask, *Do I have to know this for the test?*, or else say thanks and walk away.

Ontogeny is the development of the individual, the acquisition of adult traits and completion of the life cycle. For the young scientist, adult traits are acquired early—the asking of questions, the search for natural explanations rather than answers from authority. A kid at school said that if you put a big moth with thin antennae inside the window screen, then another one, even bigger, with big fuzzy antennae, will be there on the outside, in the morning. Thousands who have walked through my classes repeat that claim as fact. With each repetition, the fact becomes more ingrained in their minds. The people who walk into my laboratory, however, are the ones who, when told that assertion, said "really?"; then proceeded to search for their own moths to put inside their own window screens.

Modern biology is supposed to have rendered certain approaches both to science and to the training of advanced students, obsolete. We're told in a thousand ways that real science demands experimental work, that if you don't have a big grant you can't be doing research, and that the mentoring system is archaic. Some institutions cycle graduate students through a series of laboratories, making sure these students acquire the right techniques. Others demand a common coursework experience to accomplish the same thing. The logic must go something like this: If you learn *how* to do something, then that knowledge will inspire you to *want* to do something. I thought all these years it worked the other way around.

At a more ethereal level, the word "molecular" can be spray-painted on almost anything that needs to be legitimatized. There are two driving forces behind this phenomenon. First is the age old assumption that if you take something apart and study the parts in enough detail, then you'll know how the whole item works. In science, this view is called "reductionist"; reductionist biology, no matter how complex, is relatively easy to do because the materials you work with usually can be controlled—grown, cultured, manipulated in the lab, disassembled using established techniques. The second driving force is money; reductionist science, because it is based on control, often produces results that enable humans to exert even greater control over certain phenomena, and the power to control organisms such as food plants and disease-causing organisms is of economic importance. Consequently, faculty members are hired "in areas that are currently receiving funding." Parasitologists see instantly what is happening to American higher education: The teachers have become infected with the accountants' disease.

The field experience is not consistent with these so-called modern views. The Cedar Point Biological Station, the place we call Camelot, three miles down the road from Dunwoody Pond, attracts some of the most promising and talented of our young minds. Most biology majors, even those who come to the field, turn out to be health care professionals, or at least intend to. "The field," however, is typically their first encounter with serious dirt, e.g., a cattail marsh. It's also their first encounter with exotica, especially the microscopic exotica that resides in virtually every muckhole in the world. For the first time in their academic lives these people are faced with the problem of determining what a plant or an animal *is*, and of having to use highly technical literature to make that determination. Most tellingly, the real power to direct their immediate activity shifts from the teacher to the unknown world. The teacher is no longer an authority, but a fellow traveler on this exploratory journey.

Even though the problem is a tiny one—e.g., the identification of a mayfly larva—the blunt truth is that for the average well-educated high grade point average pre-med, this problem is an extraordinarily difficult one. Suddenly the value and validity of *information* are matters of extreme importance. You can't even begin to answer a simple question about diversity of life in a spring-fed Sandhills stream unless you figure out what that mayfly is. And there are no machines to help with this task. Furthermore, when you finally finish with the mayfly, there's the bucket of damselfly naiads, small crustaceans, caddis fly larva, and beetles that await their turns under your microscope, and if your experience with the mayfly is any indication of the size of this task, then it's going to be a long, long night in the lab. By dawn, the meaning of the verb "learn" will be altered; by breakfast, you'll be forever suspicious of decisions made in a climate of ignorance; and by the next evening, you'll approach the bucket from the Nevens well tank with an additional measure of maturity and a reasonable assurance that even the authorities make mistakes. Your seemingly simple exercise in comparative diversity of life in a stream vs. life in a well tank teaches you more about information in general than about arthropods.

Students never walk out on a bucket of gunk, especially if they've gotten dirty collecting it. Among the millions of species that occupy planet Earth, only one takes showers on purpose, dresses up in nice clothes, and sleeps in a warm clean bed inside a building. Furthermore, not all that species' population exhibits such behavior. We consider unfortunate those who do not, by virtue of circumstance, have access to decent shelter and clothing. I consider unfortunate those, who by virtue of choice, deny themselves access to the places that other species live. The young people described in the beginning of this chapter were invited to the field station, some of them time and time again.

They smiled and said they had other plans. I smiled back and told them to think it over. I don't believe, down deep, that they had other plans; instead, I think somebody else had made other plans for them.

6
Phylogeny

Life cycles are not assembled de novo.
— D A N B R O O K S A N D
D E B M c L E N N A N

In the summer of 1967 I drove my family to Tucson, Arizona, for some scientific meetings. Our children were quite young but they enjoyed the sights, especially the Sonoran Desert Museum. I had an old car, as usual, and got a speeding ticket in Winslow. We went across the border to Nogales to eat dinner, accompanied by the person who had supervised my doctoral research—Dr. J. T. Self, professor of zoology at the University of Oklahoma—and had a wonderful evening. The mariachi band came by while we were eating. I paid $5 for their rendition of "Malaqueña Salerosa"; the music was worth every penny and I imagined that the musicians appreciated my request for the haunting folk song instead of "La Cucaracha." Driving back north to our motel, the cactus seemed to be edging up to the highway, like wary residents of a fragile and sacred land, wondering who was venturing out across their stickery realm.

Attending these meetings was another scientist, Dr. Robert E. Kuntz, a person who had traveled widely, published extensively, and spent most of his career as a parasitologist with the U.S. Navy, serving with Naval Medical Research Unit No. 2 in Taiwan. Bob Kuntz was the first graduate student J. T. Self had ever advised and by 1964, he had retired from the Navy and taken a position with the Southwest Foundation for Biomedical Research in San Antonio, Texas. The day after our night in Nogales, all three of us—Dr. Self, Bob Kuntz, and I—were standing outside a meeting room at the University of Arizona, when Bob said, "I'm going out to see Aute Richards today."

Dr. Self agreed that they needed to visit the Richards. My ears perked up. The zoology building at the University of Oklahoma was named Richards Hall. Aute Richards had taught my own father introductory zoology five years before I was born. Richards had also supervised J. T. Self's doctoral work. I asked if I could go along to meet the man who had been my biological father's zoology teacher, my academic father's zoology teacher, and had been honored by having a building named after him, a building in which I myself had taken zoology and done my graduate research.

That afternoon we drove out into the suburbs, through some curving streets where the lawns consisted of rock, desert pavement, and cactus to a pink stucco house. On the driveway, the August Tucson heat struck my forehead like a blast from the doors of Hades. But inside the house, the living room was cool and dark, and filled with the trappings of a long life spent studying nature, reading, traveling, collecting, visiting museums and art galleries, and walking the open markets. Two very old and infirm people, Dr. and Mrs. Richards, one sitting on a couch covered with Mexican blankets, the other resting in a rocking chair, were delighted to see us and carried on a lively conversation. So this is the man who taught my father zoology, I thought, trying to envision them both as they looked and carried themselves in the depths of the Great Depression. My father earned his school money by playing the clarinet in a dance band; he had to have been starving. Out of school, on his first job as a geologist, he was paid $150 a month and had to borrow a car to carry home $2 worth of groceries. Dr. and Mrs. Richards must have been equally destitute, but here they sat, surrounded by mementos of an active and intellectually rich life, in the company of their academic child, J. T. Self, and grandchildren, Bob Kuntz and me.

I found myself trying to estimate how many students like my father and me Professors Richards and Self had encountered in their long and successful careers. During my twenty-seven years at the University of Nebraska I have given grades to about eleven thousand young people. In the period from 1920 to 1960, classes were much smaller than they are today; nevertheless, between the three of us, we must have dealt with close to thirty thousand budding doctors, medical technologists, physical therapists, nurses, pharmacists, elementary and secondary teachers, journalists, businessmen and women, farmers and ranchers, housewives, used car salesmen, insurance executives, politicians, and stockbrokers. After a few years in the profession, teachers begin to sense the impact they are having on their society. People they remember as freshman students appear, a shockingly short time later, as doctors. Two of our own children have had major dental surgery under full anesthetic, by a man who, at the age of eighteen, received an *A* in my introductory zoology course. At the other end of the economic spectrum teachers encounter their former students as waiters and bartenders; my wife smiles, out on one of our infrequent dinner dates, whenever the person who's serving us turns out to be a name and a face I can't remember from a class three years back, and reminds me of the fact. Regardless of their grades, our waiters or waitresses are usually cordial, more in control of the situation in the restaurant than in that large auditorium where I dished up full course arcanity with a quiz for dessert.

I always enjoy the encounters with former students. But the sobering statistic is how very few of them end up being professional teachers and scientists. Out of the thousands I've given grades to, I can name virtually all of the relative handful who've become practicing scholars. If my sample is representative of our higher education enterprise in general, then the proportion of American elementary school children that ends up publishing the results of original scientific research as adults, is vanishingly small and getting smaller. For anyone who's even halfway conversant with the theories of and evidence for evolutionary change, this rather obvious social phenomenon has some rather frightening aspects. A nation, and world, whose human-caused problems are simply staggering cannot survive by reducing its supply of human-caused solutions. And so far, only the scientists seem to act as if they truly recognize the long term implications of scientific ignorance on a national and global scale.

Even a little knowledge of evolutionary history and process makes one look not at what an entity is, but at what it has the potential to become and what it might actually become, given the nature of the environment in which it resides. Students, as well as problems, are of course, entities. Thus I've entitled this chapter "Phylogeny" for two reasons. First, there is a phylogeny, i.e., an evolutionary history of scientific scholars, and that history can be traced backward in time, revealing the lines of thought and selective forces that shape a nation's scientific resources. And secondly, there is a broader cultural evolution in which science plays a role, thriving or becoming endangered as the intellectual environment allows. Both of these processes share some characteristics with organic evolution. Both are relatively slow; seemingly small innovations can, in a later and different context, lead to major structural and functional changes; historical constraints, i.e., boundary conditions imposed by existing properties, force conservatism on the processes of change; and, it's very easy to draw false conclusions from cursory studies based on recent or current situations.

Such similarities between evolution in general and the scientific resource production system in particular make us look to successful teaching endeavors as models for society as a whole. What do successful operations actually *do*? How are they organized? What dynamic forces direct their actions? What pitfalls do they avoid? The people mentioned in the foreword of this book, the students who walk into my lab seeking intellectual challenges, know they are participants in a long and venerable tradition of parasitological teaching and research at the University of Nebraska. For the past twenty years, however, they've also had access to the Cedar Point Biological Station, our Camelot. The source of this nickname is obvious. "Camelot" is a metaphor referring to the inordinate richness, idealism, and freedom that surround people who go there. The richness is of biological materials; the idealism is that of people who study nature in order to satisfy a mental hunger; the freedom is that of continuous access to the swamps and prairies, the freedom from schedules and demands, the freedom to watch the stars all night then revel in the stunning beauty of a foggy sunrise over the lake. "Camelot" refers to the nobility of the work known as field biology. "Camelot" also refers to the fragility of a field station, its vulnerability to capricious decisions, self-serving behavior, actions of the politico-accountancy mind set.

As is the case with all field stations, a disproportionate number of the nation's field biologists passes through Cedar Point, some of them as established scholars, others only as unformed bits of talent. In his best selling book, *Lives of a Cell*, Lewis Thomas sang the praises of the Woods Hole Marine Biology Laboratory and demonstrated clearly, in a few short paragraphs, how MBL had played an instrumental role in the careers of many of the nation's top scientists. In Thomas' words, MBL functions as a "National Biological Laboratory without being officially designated (or financed) as such." Cedar Point is closer to a prairie pothole than a national laboratory but it, too, attracts a seemingly disorganized array of organisms—professional scientists and serious students—who migrate out of the skies over Denver, or take the short detour off an I-80 cross country trek, in order to wallow in the concentrated richness of places like the Nevens well tank. The only television set on the grounds is attached to a video camera that peers through a microscope. Dust on the road means someone's coming home from the prairie or the river. Anything wet satisfies the dress code but if it's also dirty and torn, so much the better. And if you're wet and muddy standing in the chow line, then you're making the ultimate statement.

This picture of the biologist immersed, literally, in his or her materials corresponds to the lone astronomer bundled against the cold, sitting at the mountaintop telescope, the lone chemist staring out through protective goggles into a forest of distillation columns, or the geologist isolated high against a roadcut, pounding into the sill face with a hand pick. The wet jeans symbolize the role water plays as the cytoplasmic solvent. The mud symbolizes the inorganic matrix life organized into knees, elbows, houses, cities, and ideas. And the disdain for dinner dress sends a clear message: My research is more important than my image.

The message corresponds to a mutation. Once that altered message enters the population of thoughts swirling through your head, then you are forever set upon a road that leads further and further into the unknown. Each step carries you into an ever changing realm occupied mainly by questions. Each corner obliterates the landmarks by which you might return to civilization. And at Camelot, the physical and mental worlds merge into one. Insect net in hand, fifty yards from my lab, a hundred yards from my cabin door, I stalk the shoreline weeds for civil bluets—*Enallagma civile*—Sarah's damselflies, Tami's damselflies, Aris' damselflies—the same insects that caught the minds of those who caught them. Even as I walk, and occasionally flick the net, pausing to extract the delicate thin blue animal and slip it into the same gallon plastic Tami used at Dunwoody's Pond, the unanswered questions my former students left behind also dart through the cattails, rest on a stone, or fly far out over the water where they're safe, for the moment, from a scientist's entangling curiosity. How to generate this way of being—that is the teacher's question. Take your children to the field—that is the teacher's answer; give them a net, a pan, a seine, and a lens, then surround them with people who know how to use such tools to find difficult puzzles that can occupy one's thoughts for months.

Just like erstwhile college kids now extracting DNA from mosquitoes, isolating cells from immunized inbred mice, getting on an airplane to collect sharks off La Paz, generating theoretical decision-making software, breaking parasite membranes into component lipids, or sitting in a Washington, D.C., office deciding whether to fund your research grant, I started my career as a parasitologist at a field station. The site was UOBS, the University of Oklahoma Biological Station at Willis, on the north shore of Lake Texoma. The course was taught by the person who eventually became my mentor, J. T. Self. We caught fish and small rodents and cut them up looking for parasites. The fact that one species would "choose" to live in or on another seemed like the end result of a rather remarkable set of evolutionary events. I was enormously impressed with the chanciness of the encounters that led to infection, and equally impressed by the commonness of this lifestyle. The combination of apparent gamble and obvious success told me that the bodies of animals made up a rich and easily occupied resource, but that you had to be an opportunist to take advantage of something as ephemeral as a mouse, to see the tiny rodent as a home instead of a meal. Or, more properly for parasites, to see it as a home that also supplied your meals.

In the years since those days at UOBS, I've seen this model of a student in the field, a grown-up child satisfying his or her curiosity, produce scientist after scientist. But most significantly, the products have a set of properties not easily acquired under other circumstances: a sense of the comparative approach to biology, a true appreciation for diversity especially as it is manifested at various hierarchical levels, and an understanding of the extent to which our work is shaped by our materials. This last constraint puts us in the same league as artists. The lives of wild animals are no more easily deciphered than is the skittery behavior of watercolors. How they live, what properties permit their transitions from egg to reproducing adult, and why they came to have such traits and live in such fashion, are all puzzles that together demand a holistic approach to biology, a patience with mistakes, and daily encounters with exotic conundrums.

"How" and "What" questions, however, are proximal and descriptive ones, questions of structure and function that provide information essential for the answering of "Why" questions. The latter are evolutionary questions; and when the animals are parasites, then their phylogeny is mixed up with that of their hosts, making their geneology particularly challenging to unravel. In this regard, the animals resemble those who study them; both sets of lives are intertwined with the environments that support them, but in the cases of parasites and students, the environments are also individual organisms—damselflies and major professors, the *Enallagma civile*s and Aute Richardses of the world, with their halos of influence that nurture minds bent on discovery. Fittingly, the most ardent and successful parasitologist pursuing phylogenetic questions today is a man named Dan Brooks, a third-generation intellectual descendant of H. B. Ward, who founded the American school of parasitology a century ago, started the *Journal of Parasitology*, and became the first president of the American Society of Parasitologists. We don't think of such descendants as parasites, yet we all are living off traditions born in the minds and labs of our predecessors, and in the United States, most of us who study parasites look backward and see Henry Baldwin Ward.

Ward was the first Dean of the School of Medicine at the University of Nebraska and he taught the first parasitology course in the United States, also at the University of Nebraska. He'd received his undergraduate degree at Williams College in Massachusetts, taught high school for a couple of years, then gone to Europe for graduate study. At the University of Leipzig, he came under the influence of Rudolph Leuckart, who had founded a laboratory of parasitology. Like the long parade of young scientists to follow, H. B. Ward decided to become a zoologist upon encountering an impressive teacher. Henry Ward returned to the United States in 1890, entered graduate school at Harvard, wrote his doctoral dissertation on the marine worm, *Nectonema agile*, and in 1892 became an instructor in zoology at the University of Michigan. A year later he left Michigan for the Great American Desert, also known more recently, but no more perceptively, as the Buffalo Commons.

Descriptions of H. B. Ward characterize him as domineering, aggressive, broad-shouldered, brilliant, well educated, cultured, charming, and politically adept; photographs reveal a thrusting jaw, furrowed forehead, and piercing eyes. Such types have a way of making enemies, regardless of whether they are parasitologists at an ag school in the hinterlands or commanders-in-chief of great armies. Ward spent many summers at Douglas Lake, the historic University of Michigan Biological Station, where he amassed an amazing reference collection and library. Ward died in 1945; so powerful was his imprint that twenty years later the Director of UMBS ordered all his belongings, including over a hundred boxes of extraordinarily valuable specimens, destroyed. When the UMBS caretaker refused to do the dirty deed, the Director himself personally took H. B. Ward's collections and records to the landfill, poured kerosene over them, set the fire, then buried the entire mess, personally driving the bulldozer himself. The caretaker did manage to salvage Ward's canoe paddle, which later fell into the hands of another young scientist who'd gone to UMBS to teach parasitology. The paddle now rests where H. B. Ward embarked on his professional career, at the University of Nebraska, in the museum. H. B. Ward's wife, the former Harriet Blair, was a member of the School of Music faculty and wore flowered dresses when she sang entertainment for the turn-of-the-century faculty wives. The Wards had two children, daughters.

The late 1800s and early 1900s were heady times at old NU. Isolated far out on the plains, a group of local citizens decided to form an art appreciation society, called the Haydon Art Club, which eventually borrowed its first painting, *Parable of the Wise and Foolish Virgins*, by Karl Theodor von Piloty, from the Metropolitan Museum of Art in New York City. The Club exhibited these virgins in the post office in downtown Lincoln, an event whose success led to more ambitious efforts culminating in the purchase of Leonard Ochtman's *Evening on the Mianus Bridge.* Ochtman's *Evening* became the first of a long list of major art works acquired by the Haydon Art Club and its descendant, the Nebraska Art Association. Today, seven days a week, a kid off the street can wander into a Philip Johnson building assembled from Italian travertine—the Sheldon Memorial Art Gallery—on the University of Nebraska campus, there to study at leisure Mark Rothko, Hans Hoffman, Georgia O'Keefe, Alexander Calder, Franz Kline, Edward Hopper, and a host of others who have shaped and recorded a nation's visions of itself. The path to the Gallery, like that to the football stadium and the natural history museum, is lined with sculptures by David Smith, Mark Di Suvero, Richard Serra, and Michael Heizer. Through hard, continuous, work, yes, but more importantly through idealism and vision handed from generation to generation, that small knot of artsy citizens produced a cultural phenomenon whose global and historical importance far outshine that of any local business or industry.

In its own way, the contemporary turn-of-the-century Ward laboratory functioned in a manner similar to that of the Haydon Art Club. In both cases, the driving forces were human decisions—what to study and how to study it—carried out in manners dictated in part by personal interest, in part by the nature of the materials themselves, and in part by idealism. And in both cases, the materials ultimately played a decisive role in the development of the respective enterprises. Yes, parasites, especially worms, are as captivating to some minds as paintings are to others. Harold Manter was a graduate of Bates College in Lewiston, Maine; in 1921, he got caught and decided to enter the intellectual environment generated by Henry Baldwin Ward, who by that time had moved from the University of Nebraska to the University of Illinois. When Manter left Ward's lab, he carried with him a global vision unsullied by economic considerations, eventually landing back on the prairies where Ward himself had started—Lincoln, Nebraska. From this unlikely port, Manter sailed the oceans for the next forty years, collecting and classifying worms from marine fishes. At the end of this time, he summarized his life's work in three generalizations, or as they are now called, Manter's Rules:

1). Parasites evolve more slowly than their hosts [evolve];

2). The longer the association with a host group, the more pronounced the specificity exhibited by the parasite group; and

3). A host species harbors the largest number of parasite species in the area where it has resided the longest; therefore, if the same host species or closely related host species have disjunct distributions and exhibit similar parasite faunas, the areas must have been connected in the past.

The pleasures, benefits, and rewards of an intellectual life are contained in those rules as surely as are the fates of evolving systems. Artists, architects, writers, composers, and scientists not only are privileged to have access to the material from which to generalize so strongly, they also are expected to spell out, succinctly and authoritatively, for all to read and understand and perhaps even use and enjoy, their life's accomplishments. And, they are expected to pass along the inclination to do the same.

In 1945, a Ms. Mary Lou Hanson enrolled in Biology 1 at the University of Nebraska under Harold Manter, and with that act began the process of obtaining four college degrees at once: accounting, economics, French, and zoology. She quickly evolved from a coed whose "brother could chase me around the block with a dead grasshopper" to a graduate student in zoology. Another mutation; yet another case of a student caught by worms. Ms. Hanson stayed through a year studying worms, then through another year, and another, until years turned into decades which turned into quarter centuries and one quarter century turned into two. Along the way, she passed Manter's (Ward's) mental mutations on to a parade of others. There was a lot of work to be done on the taxonomy of parasitic worms. There was always someone who wanted to do such work. And, there was always someone—first Ward, then Manter, then Mary Lou Hanson (now married) Pritchard—who would point to a table and a chair and say, "Here's your place," meaning—as understood by both—not only here's your place in this lab, but also here's your place in the world. Thus one day there entered into the suite of rooms now known as "the Manter lab," a young person about like all the other young people in these pages—raw material for the mill of cultural evolution—Daniel R. Brooks.

Dan decided to study the worms that live in frogs. The first question he asked was: What kinds of worms live in frogs around here? That question took him three years and twelve hundred frogs to answer. After he published his answer, nobody came rushing up to lay the Nobel Prize on him, no patent attorneys lined up at his door, no university presidents bragged about his work during the institution's alloted halftime minute of nationally televised football games. Dan left the prairies and went off to ask some more questions about worms, but by this time the evolutionary wheels were set in motion. The questions evolved quickly. First, the "around here" part became "continent" or "region." Dan was no longer concerned with what worms lived in Nebraska, but, in a manner reminiscent of Harold Manter himself, with what worms lived in various regions of the world, whether they were restricted to those regions, and whether their closest relatives lived nearby.

The three years evolved, just as quickly, into 300 million. The time scale over which Dan's questions were displayed became dictated not by the rules and regulations governing the receipt of graduate degrees, but by those governing the movements of continents and the fates of animals riding on crustal plates. And finally, the "kinds" and "frogs" part of his question evolved into multiple pairs of "kinds" and———, radiating out into kinds of worms in crocodiles, kinds of worms in freshwater stingrays, kinds of worms in any group of interesting hosts, "interesting" by now having evolved into a label that meant "tied somehow to the geological history of the planet."

And eventually, if not inevitably, the questions asked by Dan Brooks, the questions that started their lives as, "What kinds of worms live in the frogs around here?" evolved into something completely new and essentially unrecognizable. How, he asked, do I make sense out of all this information? Only if you'd watched the evolution occur, watched the publication record flow from a single person's mind, would you understand the connections between simple, beginning, almost elementary and trivial questions about the worms of Nebraska frogs, and the elegant, all-encompassing, highly general and transferable questions about whole groups of species riding continents and drifting away from their relatives for millions of years. A hundred scientific papers and several books later, Dan Brooks took to the podium at the national meetings of the American Society of Parasitologists to accept the highest honor his scientific society could bestow: The H. B. Ward Medal. The circle was complete, some said, and fittingly so. The phylogeny is revealed, said others, just as fittingly. This is where scientists come from: They are shaped by traditions, opportunities, encouragements, a place to get your hands dirty, and great big problems. Look for those features in a landscape, and that's a center of origin.

My memories of the visit to Aute Richards prompted a letter to my own former adviser, J. T. Self, in search of my own evolutionary history, although I waited twenty-six years to write it. The purpose of this letter was to explore the validity of the above assertion about the way in which our human resources are developed. What kind of a working environment had Richards provided? How had that environment been passed along to the next generation of major professors? Not surprisingly, Dr. Self's recollections of our afternoon in Tucson were tinged somewhat with his own impressions of student life. Self remembered the Richardses' quizzing of Bob Kuntz about his Navy career, his duties in Africa and Taiwan, what he was going to do in his retirement, and about the book most of us were waiting for Bob Kuntz to write. In the late 1930s, Kuntz was a kid who, according to JTS, "would forget to appear for his report in cytology because he found some tadpoles more interesting." But in the ensuing years, Bob had become a productive scientist, an unmitigated field naturalist whose collections—tapeworms, nematodes, spiny-headed worms, pentastomes (enigmatic arthropod-like parasites of snakes, typically poisonous snakes)—had bolstered the careers of many to whom he'd sent specimens from the far corners of the globe.

Aute Richards, ever the teacher, saw Bob Kuntz' career as a parasitologist in the U.S. Navy as serious book material. There seemed to be a reason for Richards' needling on this point; duties in Africa, Taiwan, and Bethesda, Maryland, had taken former student Kuntz to some of the world's parasitological hot spots, and he represented a unique combination of the military's recognition of parasitism as an enemy, along with his extraordinary breadth of interest, a driving curiosity, and seemingly boundless energy. Kuntz demonstrated, and Richards the teacher knew it, that jobs are obtained, but careers are lived and intellectual accomplishments are built, piece by piece laid in place by individuals, regardless of, and occasionally in spite of, jobs and careers. Richards had never forgotten the evolutionary principles as they apply to the phylogeny of scientists: Patience, tolerance, dignity, and opportunity are the key elements of a fertile garden. Richards' pressure on Kuntz to write the book was his own way of telling Bob that society must not be allowed to forget how scientists grow best and bear fruit. But Bob Kuntz also remembers the fine touches Richards sought to put on his students.

"He showed us how to make round dots," says Kuntz, describing Richards' teaching techniques. I still have my father's undergraduate zoology notebooks from the early 1930s and can attest to the fact that when it came to shading your drawing of a specimen, Richards encouraged dots of perfection. "He always wore his Phi Beta Kappa key on a gold chain along with a magnifying glass," Bob continues. "He'd come around and check your dots with a magnifying glass and get pretty upset if they weren't round!" The symbolism of round dots was not lost on Bob Kuntz, who at the time was working weekends in a distant town and collecting his research animals on the side. Forever after, he paid full attention to the finest details of his work and treated his biological materials with careful respect. Richards' pride in Bob Kuntz' career was that of a teacher who demanded round dots, but saw the most potential in a student who wasn't intimidated by a teacher who demanded round dots and checked them with a hand lens.

Fittingly, J. T. Self's memories of the Tucson visit were characterized by two elements of the teacher-student relationship: the acknowledgment that a student was engaged in a worthwhile pursuit, and pride in that student's accomplishments. John Teague Self spent the first twenty years of his life as a "cotton picker and weed hoer," entering graduate work at the University of Oklahoma "green as a gourd," according to his own father, and with a level of social development he characterizes as "O." Aute Richards, his adviser to be, rested at the other end of the academic, social, and political scale. The Richardses loved to entertain, regularly laid out their best silver and china, and graced the occasions with their formality—Richardses carving the roast, Mrs. Richards serving the vegetables, then returning the plate to her husband to distribute, alternatively, to the right and left. Self is a bear of a man. He could not have been much less of a bear of a young strapping graduate student when he was first invited to one of these dinner parties. The table must have been crowded; Self, the left hander, felt compelled to eat with his right. The switch made him painfully aware of his need to acquire some social graces in order to mix with the gentle and traditional, although somewhat impoverished, grace of academia. Ever the teacher, Richards helped.

"Teague," he asked one day a week after an elegant evening, "did you enjoy the party at our house?"

Of course the young green-as-a-gourd former cotton picker and weed hoer now turned biologist admitted that he'd enjoyed it immensely.

"Then go into my office," said Richards, "and call Mrs. Richards and tell her so."

Decades later, the Selfs stepped into the Richardses' role, inviting a laboratory full of graduate students, along with their wives and husbands, to their home for dinner. I didn't consider myself a social zero, but evidently was closer to that value than I thought. Mrs. Self made chili that could have won any contest. My first encounter with this heavenly food is recorded for posterity on the recipe card she eventually gave to my wife Karen. At the bottom of her list of secret ingredients and well-practiced preparative hints was the note that commemorated my immature affront to her culinary virtuosity: Get table catsup for John Janovy. I'd neither understood nor appreciated the subtle artistry involved in making chili. Now, thirty years or so later, usually when the first truly miserable cold fall weekend settles in on the northern plains, Karen pulls out her recipe card and reconstructs that brew. And in reenactment of an event I can never remember but she can never forget, she puts the catsup on the table in front of me.

Ida Self's chili is a metaphor for the intellectual soup that instinctive teachers prepare for their students. And when those students are biologists, then a field station is the key spice. Dunwoody Pond may be a source of worms and snails, but perhaps more importantly it's also a source of humility, patience, and organic dirt the Ancients thought spawned flies and toads. A tiny fraction of the planet, isolated deep in a relatively depauperate ecosystem, Dunwoody Pond nevertheless serves up awareness: of your own ignorance, of how much there is to study, of problems that someone could work on if only that someone could be diverted out of the vast river of human resources flowing into chasms labeled "careers" but looking suspiciously like day jobs.

I guess it all comes down to money, doesn't it? Pure and simple money. Rent checks. Sacks of groceries. These are the items that screw up the chili. Where can I get a job as a taxonomist who studies the parasites of damselflies? Nowhere. There is no such job. Like Bob Kuntz, you have to do another job in order to study the parasites of damselflies. But like Bob Kuntz, Harold Manter, J. T. Self, Mary Hanson Pritchard, and Dan Brooks, if you can find that place where it's all right to study something difficult that no one else seems to think is important, then you will evolve into a biologist. Eventually, if that evolution proceeds far enough, you'll find yourself in their position, faced with a left-handed upstart who eats with his right, or a right-handed upstart who asks for catsup. You'll smile, and put that store-bought bottle on the table, knowing that inevitably, if you've followed your chili recipe correctly, then there'll come a time when the kid won't ask for it anymore.

7

The Swallow Barn

*. . . our Western science—and there
seems to be no other—did not start
with collecting observations of
oranges, but with bold theories about
the world.*

— K ARL P OPPER

If someone should ask: Where is the laboratory in
which important scientific work is being done? I would probably
reply with another question: What do you mean by "important"?
Oh, you know, he'd say, work that is published in the most respected journals in the world, work that seems to reshape the
way we view our own lives as well as those of others, that sort of
thing. Then I'd say, get in the old station wagon, I'll take you
there. We'd drive out the west gate—The White Gate—of Cedar
Point, turn to the right past the hydroelectric plant and follow
the washboard road beneath the dam. When we got to the black-

top Keystone Road, we'd turn right, go past Tom McGinley's campground, store, and laundry, past the State Game and Parks campground, and stay on that road until we could see the original McGinley family ranch headquarters. Then I'd slow down, pull over onto the sloping shoulder as far as I safely could, and park beside a concrete box culvert through which runs a small stream. Here we are, I'd say. My guess is that I'd get a stare of disbelief in return.

We've arrived at the intersection between Whitetail Creek and a road with no name other than "the Keystone Road," which by local agreement means the road *to* Keystone *from* the big dam, the road you have to go on to get to Charles Brown's laboratory if you go out the west gate. You can get to both the lab and Keystone by going out the east gate, too, but that road is called "the Roscoe Road," mainly because Roscoe is at the southern-most end of it, where a scenic and curvaceous descent from the outcrop canyons along the North Platte River meets State Highway 30. The Roscoe and Keystone roads cross north of the North Platte River bridge; the Roscoe Road pavement ends there. Dunwoody's pond is a mile north and half a mile east of the pavement's end. The lab where all this important scientific work is done, i.e., the Whitetail Creek box culvert, is a mile-plus west of that intersection.

A large white cylinder now marks the spot; a gauging station has been installed in the mouth of the culvert south of the road. This seemingly incongruous intrusion upon the otherwise sandy green landscape evidently has had no negative effect on the lab's operations, or on the hundreds of cliff swallows that in a typical year use the Whitetail culvert as a nesting site. And like his beloved swallows, Charles Brown migrates over several thousand miles annually, returning usually in April, to the Whitetail Creek culvert. Charles is a highly focused, somewhat lanky, bearded, and often field-dressed man whose Texas accent and Dallas Cowboys bumper sticker expose his roots; he's also the person who has converted the culvert into a globally recognized laboratory without altering a single rock, grassblade, piece of concrete, or installing an electrical outlet.

"Whitetail Creek." Charles doesn't hesitate even a fraction of a second in answering the question: Where is your most important site? "At least since 1982. It had one nest in 1982 and peaked at twenty-three hundred nests in 1990. We've made all of our most significant discoveries there. We've worked there the most and the longest time. It's one of the biggest of all our colonies, the most accessible. We've banded the most birds there."

In that short paragraph, he summarizes the role that one spring fed stream and a concrete box culvert have played in the development of a scientific career that began in his backyard, progressed eventually to Princeton for a Ph.D. and then to Yale, where he's now a young professor. It's also a site that has played a major role in the development of young talent that comes to work every summer on the swallow project. The places where a field biologist builds a career and teaches others to do the same become a part of your inner self. That fact is not lost on Charles.

"I have an emotional attachment to that site," he continues. "It's clearly my favorite." Then, recognizing his assistants' contributions, he adds, "But it's not necessarily the help's favorite site. It's been very busy; there are a lot of birds to catch, so you seldom have time to sit and appreciate it, especially during the nesting season."

This culvert also plays a major role in the publications emanating from Charles' career, papers that have shaped, to a large degree, our perceptions of life in large groups. The ideas, the research, the animals, and the culvert are merged into a single phenomenon—a practicing scientist and his official record.

Charles Brown goes to western Nebraska in three pickup trucks, with Mary—his first research assistant in 1982, now in her twelfth year not only in that role but also as Mrs. Brown—a small crew of additional assistants, and a gigantic pile of equipment characterized mainly by its simplicity. The most complex of all his stuff is the computer: first on the list of items scientists now take to the field, but did not take a generation ago when pencil and paper sufficed to record the results of our classic experiments. But everything he takes, as well as that which he stores locally during the winter, has one function, namely to make possible an observation.

"Observation." He responds instantly when asked about his investigative tools. "It's the major tool used throughout this study, either observing the birds' behavior, or where they are. We've done some experimentation, but always in a field setting. We've not manipulated the system to any great extent." We talk about the finances involved in his research; he enumerates the items essential to his research on a day to day, even hour to hour, basis. "Lawn chairs are very important pieces of equipment. Army shovels. Waders and rubber boots. Poles—conduits. Dirt. Plastic buckets to put the dirt and net poles in. Ectoparasite sampling jars, wide-mouth jars with a little hole cut in the lid for the bird's head to stick through. Ether to anesthetize bugs, fumigant, Tanglefoot, radio transmitters, sprayers. We glue the transmitters to the back of the bird, to the skin; they stay on up to a month. Banding pliers, which are an incredibly obscure piece of equipment but without which we couldn't work. A scale, a little spring scale, bird-holding bags—mesh bags—binoculars, paint pens to color-mark foreheads." He pauses. "And bands. The bands are free."

The bands are free if you have the permits to use them. The permits are free, too, if you have the expertise to justify their issue. The expertise is not free; it's paid for in a thousand ways, including cash. I ask for an estimate of the total amount of money he's spent on equipment in the last ten years.

"Except for vehicles and computers, we've spent maybe five thousand dollars."

"That's a lot of research for five thousand dollars."

"That doesn't include gasoline." Nor time. Time and transportation are to an ecologist what giant telescopes are to astronomers. Time and travel are included in the price one pays for expertise. I suppose an accountant could, and might be inclined to, put a price on Charles Brown's expertise, or on the expertise of thousands of scientists who work at fundamental, but unprofitable, endeavors. It is at this point that economics reaches an impasse with the human spirit. One cannot, in the final analysis, put a price on Charles' head without reducing all individual, sustained, high-level, intellectual efforts to simple commodities. At some point in his listing of cash purchases he reaches this impasse. He can tell you the cost of a shovel and a bucket, but he doesn't estimate the value of his time and learning.

Charles is quite certain about one item, however: His most important pieces of equipment are the nets; without them he and his helpers could not do their work. Japanese mist nets are large, baggy curtains made of very fine, and very strong, black nylon threads. Birds fly into them, get entangled, and then, depending on the part of the world where the nets are strung, get eaten, sold, or banded. At the Whitetail Creek culvert, cliff swallows get banded, but on some occasions in the past also have had their white foreheads marked—color coded—for individual recognition. Cliff swallows are among the most colonial and social of animals; their lives are locked into that of the group. Yet, in order to understand the factors that bind this social unit together, Charles has no choice but to focus on the individual. And his first task is to catch and band birds, a task that is made difficult by political and ecological forces far beyond his control.

"Mist nets are increasingly expensive and becoming harder and harder to get. The Japanese are not exporting them, primarily because of the conservation organizations' efforts. They were being used to decimate native bird populations in many places. So there is a blanket ban on the shipping of mist nets, including to the United States, and last year you basically couldn't get new mist nets in this country. The big forty-two-by-thirty-foot net is worth about two hundred dollars. But a normal eighteen-foot one costs about twenty-four dollars.

"We're going to reach sixty thousand banded in the next four or five days." Charles is sitting in his lab in a building called The Swallow Barn, talking about nets, bird banding, and politics. "The person who bands the sixty-thousandth bird gets a cake from Connie. The cake will be in the shape of a cliff swallow." Connie Andre is the Cedar Point Biological Station cook, a job she performs as much out of love as for money. Building animal-shaped cakes to celebrate significant research accomplishments is one of her specialties.

The Swallow Barn, in which Charles sits talking about his career, is a recycled building. In the late 1930s it was used as a headquarters for the crew that built Kingsley Dam. It sat vacant and abandoned near the spillway until the early 1980s, when it was moved to a spot near the Cedar Point west gate and used as a storage shed during the construction of the hydroelectric plant. After the power plant was finished, the Central Nebraska Public Power and Irrigation District offered to give the building to the field station. In the meantime, barn swallows had built nests up under the fifty-year-old eaves. When the building was moved almost a mile east, creeping slowly down the gravel road and up onto its foundation in the hills, the swallows followed their nests, taking up residence and completing their broods in the new location. Ron Randall, the station's physical facilities supervisor, took a careful look at the building and passed judgment:

"Well," said Ron, "it needs a new roof, new windows, new doors, new floor, new ceiling, new siding, new wiring, plumbing, insulation, and air-conditioning, but other than that, it's a good building." He was looking at the heavy floor and ceiling beams. "They don't make wooden buildings the way they did before the war." He was of an age when "before the war" meant in the 1930s instead of the 1950s. Ron then took out his hammer and started to work. A few weeks later, he stood back and surveyed the fruits of his labor. "I left that strip of old siding underneath the eaves on this end," he said, pointing up toward the roof line. "Maybe the birds will nest up there again next year."

We had a building naming contest for the "new" lab. It was no contest. Charles Brown submitted the winning entry and his operation is now housed in the building he first called The Swallow Barn. Scattered on the tables in his lab are the tools of his trade: clipboards, binoculars, mesh bags for holding birds. Sitting in a circle, talking about their experiences, are Charles' assistants for the year: Mike Kostal, Audrey Hing, and Cecily Natunewicz. Charles recruits assistants from across the nation. This year they are mostly Yale students, working largely as volunteers. Their pay is the experience of spending all day, seven days a week, for weeks on end, with Brown and his birds. Mike describes the benefits of banding and netting swallows.

"Except for the cake, the only rewards are for people who catch birds from 1982—we get Dairy Queen sundaes." The Dairy Queen is ten miles away and a trip to town is a 30- or 45-minute hiatus in the otherwise constant fieldwork. A bird banded in 1982 would be eleven years old this year. I ask how many tiny swallows have lived that long, making annual migrations to the Southern Hemisphere. "There are a few," Mike says. He understands the subtle gentleness of a sundae in exchange for such geriatric evidence.

"We had three from 1982 last year," adds Charles. "We caught two this year. There are still two birds alive who were babies when this all started." Charles has actually touched two cliff swallows, small birds you can hold easily in the hollow formed by your fingers curled to touch your palm, that have flown tens of thousands of miles, through and over deserts, swamps, rivers, oceans, forests and prairies, returning eleven times from the Southern Hemisphere to a few square miles of Nebraska.

These birds are the equivalent of a 110-year-old great-great-grandmother who's just come back from her eleventh trip to the Moon; the band is a bracelet she received as a present when she was born. The symbolism is not lost on Charles Brown or on his assistants: By their labor, and the labor of the other two dozen or so people who've been drawn to this particular swallow project, these long-lived individuals have redefined the meaning of the words "cliff swallow."

But of all his accomplishments, Charles' most important one may border on a philosophical principle: In order to redefine what it means to be one of the planet's most social species, one has to redefine what it means to be an individual member of that species. To that end, he's marked—thus "individualizing"—not only individual birds by banding, but also everything else about a swallow's life, from blood-sucking parasites to colonies.

"We've individualized birds, we've individualized nests, colonies, eggs, babies, mudholes, predators." Birds have been marked in two ways, bands for life, paint for the season. The years they painted adults, Charles' crew would appear at dinner in multicolored polka-dots. Cliff swallows have a white triangular patch on their foreheads. By using several colors of paint pens, arranging the streaks in all of the possible permutations and combinations of black, red, yellow, green, purple and blue, they were able to mark hundreds of birds. And the swallows were naturally cooperative; sitting in their nests, looking out of the entrance tunnel, the birds "told" who was home when and for how long; and, if an observer were quick enough with the binoculars, he or she could find out who was in the "wrong" place—cheating or stealing.

The nests were easier to mark than the birds who occupied them. But the marks are no less permanent than an aluminum band. Still, more than a decade after Charles Brown began his studies on the swallows using local culverts and bridges, the nest markers remain. A parasitologist spends a lot of time under culverts and bridges; usually there's a convenient parking place nearby, and fish, snails, aquatic insects, and parasitic worms live under bridges as well as up and down several hundred miles of river. I always look up at the swallow nests when I'm under a bridge. It gives you a feeling of respect to see the numbers on the beams; those numbers mean a young scientist has been at work, even if that scientist is no longer exactly what you'd call young. Even the mud rings are numbered, evidence that somebody built there in the past but didn't come back to build again in the same location. I always pause to read the numbers, reconstructing the plans, approaches, hypotheses being tested, results being tabulated, papers published, read, and argued about—the intellectual history implied by those felt-tip numerals on concrete. I find myself wondering how long the nest numbers will remain visible, whether fifteen or twenty years from now some student will ask some teacher what they mean. The two will be standing in the river looking up under the bridge. The teacher will say those numbers mean that a scientist came here one time looking for the costs and benefits of living in large colonies. And the student will ask if you mean that they had to mark every single nest to find that out? And the teacher will reply yes, you can't find out what makes a society hold together, or disintegrate, unless you understand what it means to be an individual in that society.

Eggs are easiest of all to mark. Charles and his assistants simply write code numbers on them in pencil, like you'd write "hard boiled" on some chicken eggs you've put away in the refrigerator for use in a salad this weekend. Egg marking has revealed one of the big costs of living in large groups, at least in the case of cliff swallows. There may be advantages to having neighbors, but with so many neighbors, fidelity has evidently lost some of its allure.

"These animals are extremely devious," says Charles, "they employ a variety of sneaky reproductive strategies. This would include such things as giving an alarm call when no predator is present. Then when everyone has flushed out of their nests, jumping into your neighbor's nest and laying an egg. Or sometimes they'll throw out a neighbor's egg, or steal nest material, or move an egg around when their neighbor isn't there."

I wondered how they moved eggs.

"They pick them up in their bills and fly around from one nest to another." Even Charles is still amazed by the behavior of animals he's come to know better than any other human being has ever known them. "We've seen it three times." In twelve years of work and steady observation for weeks on end, he's seen swallows pick up an egg and move it to another nest three times. He pauses, thinking. An event seen three times in twelve years hardly counts as commonplace, but on the other hand, there are so many colonies and so few people to study them. A classic line from an emotional film entitled *Field of Dreams* has entered the American lexicon of cliche and now best describes the various bridges, culverts, and irrigation gates constructed by the Nebraska Public Power District. If you build a concrete wall, especially a protected one, the swallows will come, usually within a year. (Charles actually has a research site he calls *Field of Dreams*, a concrete block building in the city park, near the baseball diamond, in downtown North Platte!) A Martian-canal maze of ditches and bridges now spreads throughout a hundred-mile area of the Platte Valley and adjoining prairies. And when the birds built on these structures, Charles came. Three egg carries observed in twelve years may not be a measure of rarity so much as a measure of how much swallow behavior Charles and his team have yet to record. Like most long-term research endeavors, this one's discoveries lead mainly into deeper mysteries, telling you that no matter how much work you've done already, there is still something left unseen and unexplained. Thus, although Whitetail Creek may be Charles' favorite, a constant search for material leads him on a tour of the District's handi-work.

"Beckius Bridge," he begins when asked about his other research sites, "except that we call it IS#1—Irrigation Structure #1. We've used it sporadically since 1987; it's not used by the birds every year. We haven't done much except catch the birds. We can't reach the nests there." When Charles isn't able to reach the nests, to count and number eggs, count and weigh babies, a colony slips down on his list of important places. "There are some Ash Hollow sites," he continues, referring to the general area at the west end of Lake McConaughy. A historic place by local standards, Ash Hollow State Park now includes a cemetery whose headstones carry settlers' names, a museum built over a pre-Columbian rock shelter, and an idyllic little park. "One colony's actually at Windlass Hill." The hill is named for the windlass used to pull covered wagons up the still visible ruts south of the North Platte River crossing.

"Clary has been used every year except one. It's a very important site. We've made a number of very important discoveries there. There's another little culvert we call 'Bluffs.' " "Clary" is a large, two-channel concrete box culvert similar to the one at Whitetail Creek. Clary Ranch, at the west end of the lake, also is a bison kill site, where anthropologists reconstructed the life of Native Americans long before sociobiologists came to study the cliff swallows. The Clary culvert is a particularly memorable one for me; not often can one point to the exact place where a population's burden of parasitism is discovered and described, all within a single summer. It was there that Charles took a can of Tanglefoot, incredibly sticky gunk used by ecologists to sample insects, and painted a strip that cut the colony in half, then continued up onto the roof of the culvert and down along the walls. Swallow bugs are a form of bedbug; they are also wingless. Bugs that tried to cross the strip got stuck, like flies on flypaper. Then, using a pump sprayer straight out of the 1930s, he fumigated half the colony, ridding it of the blood-sucking parasites. The results were startling.

"Clary was the first colony where we first began to see the effects of these parasites on the birds. That's where we took the picture of the two babies of the same age, one big, the other sickly. That picture's become somewhat of a classic; it's been reproduced about ten times now. We have since done that at other colonies. But Clary gave us our first insight into the effect these parasites have."

The nestling from the fumigated portion of the colony is almost twice the size of the one from the unfumigated side. The picture is one of potential as much as effect; you realize what nestlings everywhere could achieve if only relieved of a massive burden of parasitism. The lesson is too obvious to be ignored: We are all nestlings at some point in our lives, and the swallow bug is all the forces—both mental and physical—that drain away our futures.

"Mule Deer is a culvert under the interstate between Paxton and Roscoe." Charles' list of research sites continues. At numerous times I've been driving down the interstate and glanced over to see his truck parked by the Mule Deer culvert. It looked like an interesting place for a parasitologist. "We get there by way of the access road south of Roscoe, south of I-80, then back over. It has a lot of muck and standing water, and dead things." It is an interesting place for a parasitologist.

"Canal Bridge, Keystone Bridge; actually we net at about ten or fifteen different canal bridges between here and Sutherland."

But of all the local biota, bridges, canals, nest-mud puddles that Charles and his assistants have marked, or otherwise individually recognized, the most intriguing item is a single murderous grackle.

"We had one grackle that killed seventy birds at a colony

and we know it was the same bird because he had a missing wing feather. The other grackles went by and didn't do anything. But this is an individual grackle that learned to specialize on these birds." We sit in silence for a moment, thinking about the rather obvious implications of this observation, but knowing at the same time that biologists are not supposed to anthropomorphize their behavioral observations too strongly. As scientists, we also recognize the good fortune that a murderous grackle had a missing feather and so could be identified. That single feather is the difference between condemnation of an individual and condemnation of a species. But if he were to condemn species from his work at the Cleary Bridge, Charles would choose the fleas and swallow bugs that he knows are far more deadly than grackles and bull snakes. "We have marked swallow bugs and have started mark recapture studies on swallow bugs." The audaciousness of that endeavor leaves a moment of silence again. Yet these bugs are remarkably long-lived; they can go for a year without a blood meal. They also are remarkably mobile, in the geographical sense, occurring everywhere cliff swallows occur, a feat that can be accomplished only by riding if you're a wingless bug. But as is the case with the swallows, you can't say much about colony bugs unless you know how an individual in that population lives. Thus the marks.

In the life of an individual swallow, mayhem does not end with snakes, grackles, bugs, and fleas. Charles calls it "forced copulation" but it's called rape in the newspapers. The birds are particularly vulnerable at the times and places where they gather nest mud. A rain puddle three feet across along the washed-out apron of a gravel road is a typical site. You can see the mass of birds a quarter of a mile away; they're pouring in and out of the sky; as you slow down, lifting your foot gently from the gas peddle, the dark splotch on the ground resolves itself into a mass of fluttering wings. The birds are so focused on the gathering of nest material, they're almost oblivious to your approach. Even if you scare them away, they soon return; their business is evidently more important than your threat.

It was at such sites that Charles and his assistants used a stuffed swallow mounted on a board to calculate the rate at which rape occurs at watering holes. It quickly became evident that forced copulation not only was common, but also was one of several factors that contributed to misdirected parental care and mixed-parentage broods. The birds have a distinct tendency to trespass into neighbors' nests, especially early in the season. Most of the time successful trespassers just sit there. At other times, they steal fresh mud or nest material or toss out eggs. In Charles' own technical language, egg destruction "prolongs a female's egg-laying period and thus keeps her sexually receptive to forced copulations that the egg-destroyer himself may also perpetrate." Egg destruction, transfer, and deposition in neighboring nests result in a significant number of mixed-parentage broods. Charles estimates that nearly a quarter of all nests contain eggs from more than one set of parents. When rape is added to the list of factors contributing to mixed families, the figure increases to over 40 percent.

Females are far from guiltless in this system, preferentially choosing neighboring nests with relatively low swallow bug populations in which to lay extra eggs. And adolescents add to the mix; recently fledged birds enter nests—not their own—with smaller young and take food from the returning parents. This thievery is highest among birds fledged from larger colonies, with robbers even cruising among colonies several miles apart. The intent, apparently, is theft. There is no evidence these birds are just trying to get back home. Indeed, the most immediate hangout of juvenile swallows is with others of their kind and age, often blanketing power lines in large groups—up to several hundred birds—from a number of colonies.

Charles Brown came to Nebraska to study the costs and benefits of coloniality. In watching him work for a decade, listening to his seminars, and reading his results, I've seen many costs. I ask him about the benefits.

"The major advantage for cliff swallows is the opportunity for them to observe one another finding food." I remember when Charles gave his seminar on group foraging, pointing out that birds sitting in their nests would watch their neighbors return, evidently determining which ones had foraged successfully. When those birds left again, the ones back home would follow. The whole enterprise of social foraging seems more laid back, less vicious and stressful, than that of social breeding. In contrast to their sexual deviousness, there's no indication the birds try to hide food. And the food itself is highly unpredictable. Over a several year period, you know there will be swarms of insects, you just don't know which year they will be plentiful, nor where they'll be located exactly. And the birds do eat relatively large numbers of swarming insects, especially midges. When Charles uses the term "information transfer," relating to virtually all aspects of behavior, I naturally ask about learning, especially the deviousness.

"Learning has not been looked at in these birds to the same extent as it has in some other species. But clearly these animals learn. They learn people. They respond differently to different people. They know me. Without question they respond to me more than to the average person. They exhibit alarm. They clearly know that when I'm there, something is going to happen to them. They become agitated. They also know the truck because when the truck drives up, they flip out. But if I drive up in another car, they don't respond. So they learn the blue truck and the guy with the beard and the cap and the shirt with bird droppings all over it. If I go over there cleaned up, without my cap, and in a different car, they don't respond. So, they do learn."

"Tell me about the group foraging, the following of feeders."

"It's hard to say whether that's a learned behavior, whether the bird is more likely to do that based on its initial experience." Charles pauses and thinks for a moment. Here's another question he can add to his list that comprises a lifetime's research completed, and another lifetime's research yet to do. This single human's curiosity about a single species has been successfully converted into a library of understanding about group living. Most of his observations reveal events and processes that are highly analogous to those occurring in our own species. It's tempting to extrapolate from swallows to people; but such extrapolations are frivolities best indulged in late at night watching for shooting stars and talking with tolerant friends.

Yet regardless of whether Charles' research results will ultimately "benefit mankind"—as the practicing scientist hears it phrased by the taxpayer so often—his project has produced, and continues to produce, a steady stream of young talent. Most of that talent comes seeking professional level research experience and finds Charles. This seeking and finding knows few limits, and geography is certainly no hindrance to the movement of bright students. Audrey Liang Yin Hing's journey to western Nebraska, for example, began at the Tanjong Rhu Elementary School in Singapore.

" 'Rhu' is a Maylay word for the casuarina tree—a conifer, and 'tanjong' means an area," she says. "So Tanjong Rhu is an area where there are a lot of these conifer trees." From the grade school named for trees, she progressed through a junior college named for a saint, before ending up at the National University of Singapore and writing her honors thesis on egg proteins in tilapia, a commonly cultured food fish. Molecular biology carried her to Yale. Curiosity, and a desire to "mix genetics with ecology, to look at the genes that determine an organism's interaction with its environment, from micro to macro levels" led her to The Swallow Barn and her mist net duty. Audrey had never been on the plains before. She flew to Denver and took the bus to Ogallala. She'd never been on a Greyhound Bus before, either.

"The driver was very nice; he was from Missouri; he told me his whole life history. Showed me the album of his family." Audrey Hing chuckles at the friendliness. "But he was curious about why I wanted to come out here to help a guy work on swallows when there were so many other big birds flying around." I tell her my latest Duane Dunwoody story: He wants to put a radio collar on a vulture to see where they go and where they nest. It's a lot easier to find a swallow nest than a vulture nest. That's one excellent reason for making the trek from Singapore to Keith County—material is plentiful. So are the lessons.

"I've always been interested in genetics and ecology," she continues. "I've always pursued genetics, but decided it was time to explore my other interest. Eventually I'd like to marry the two. Right now I just want to do ecology. Everything I knew about ecology I'd learned in books. But I wanted to come out to see what it was like, what the difference was between doing it and what was in the books."

"What is the difference?" I know the difference. Everyone who's gone to the field knows the difference. I was curious to find out if one could learn the difference in a few short weeks.

"A lot of routine boring stuff," she says. Yes; it doesn't take very long to learn the difference between science in books and science as it's actually done. "But it's the same in every field." Nor has her molecular biology experience been for naught. "The ideas are exciting and the data may answer your questions in interesting ways. But getting the data is sometimes boring."

Mike Kostal is a swimmer at Yale. His life up to the point of coming to Cedar Point is distance freestyle—September to March—with a couple of weeks off before he starts up again, swimming another season between April and August. He's been out of the water for two months now, handling swallows, and is "somewhat behind physically." But he's convinced his swimming will be better after this break.

"All the problems in swimming are from the neck up;

there's no such thing as physical burnout." And if the problems are in your head, what do you think about? "Stroke count, during the 1650, and where the other guys are, whether they're drafting." Swimmers draft in the same way race car drivers do. The "1650" is 1,650 yards, i.e., 110 feet short of a mile. He's not concerned with stroke count and drafting when he's banding swallows. "Catching and banding is really a complex activity. There's the activity of drop netting; there's the activity of getting them out of the net; there's the activity of processing, as when you sex and read the band numbers, or put a new band number on; and then there's recording."

"So what's on your mind when you're doing this?"

"I think about everything. I think about swimming. I think about school. We talk. Sometimes inane, sometimes serious. We talk about politics." I don't ask whether he considers his political talk inane or serious. I do ask where he'd like to be a few years down the road and what his summer off from chasing black lines down the bottom of a swimming pool has contributed to the journey.

"I'd like to get a good grounding in how the Earth's natural cycles actually work—the chemistry and physics involved, the time scale, and then see how that information applies to the effects humans have on the planet." That sounds like several lifetimes worth of labor. Mike appreciates that fact, but is not overly intimidated by it. In addition, he understands the mental component of long term, large scale, studies of natural phenomena just as well as he understands the mental component of racing a mile freestyle. "It's a nice exposure to what is obviously a rather well-planned research endeavor and to the logistics of doing it. You get to see how simple and bare bones it actually is, but also how well thought out it is, how well designed. And that's something you can apply to any field. This is the type of thing young scientists need to see and don't see in college."

Cecily Natunewicz arrived at the field station via Houston

and New Haven, having started her intellectual life as a comparative ecologist and molecular biologist at about the age of eight.

"When I was little," she says, "we lived in Wisconsin but spent the summers in New Jersey. We had a beach house. And I always got interested in the sea life. We used to pick up jellyfish. When we'd go back to Wisconsin I'd compare things on the shore of Lake Michigan with what I'd found in New Jersey. Of course it wasn't a major comparison; I was only eight years old. And, then, I took an ecology class in the seventh grade. We had a choice in the public school—geology, weather and astronomy, or ecology; I picked the last two. I always wanted to be a marine biologist. Stingrays and jellyfish are my favorite organisms."

There must be something magical about middle school; all three of Charles' current crew members have strong memories of seventh grade science. But Cecily remembers making molecules out of marshmallows and toothpicks even in the second grade.

"We ate the molecules." She chuckles at the symbolism obvious to any adult biologist: It may smell and taste like steak and ice cream, but it's still only molecules.

"What are you going to be doing in twenty years?" I always wonder what young people are going to be doing in twenty years; two decades seems like eternity to them, a flash in the pan to me.

"I'll probably be doing research somewhere. Hopefully somewhere on a large body of water. Hopefully marine ecology. I might be working with shrimp or sharks. Or even jellyfish."

"What's the attraction to jellyfish?"

"Oh, when we were in New Jersey, we'd pick them up. Some would sting and some wouldn't. You don't pick up men-of-war. Usually you don't find whole ones on the beach. Just tentacles. We drew pictures of them."

"When was the last time you drew a picture of a jellyfish?"

"Probably this spring, doodling on notes." I know the story well—sitting in a university classroom ostensibly taking lecture notes but drawing instead, pictures of animals you've handled since you were a child. As with Mike Kostal, I ask Cecily what goes through her mind as she's banding cliff swallows.

"All sorts of things. What I'm going to find on my walk around the lake tonight, what I did yesterday, friends from back home, organic chemistry." Marshmallows and toothpicks.

Swallows fly home and sleep when the sun goes down. Charles retires to his Barn, recording data. His assistants pick up books.

"*Satanic Verses*," says Audrey, when asked about her recent recreational reading, "and Steinbeck's *East of Eden*. P. G. Wodehouse is one of my favorite authors. I read plays, too, scripts, some by Tennesse Williams. Also lots of poetry." Gerard Manley Hopkins is one of her favorites. Why? "He takes a lot of liberty in creating his own words, words that just go with the feeling." Some of my strongest memories of a professional scientist at work are those of George M. Sutton, Research Professor at the University of Oklahoma, bringing a Hopkins poem to class—" The Wind Hover"—and using it as a starting point for his lecture.

"*Picture of Dorian Gray*," adds Mike, "by Oscar Wilde. My girl friend sent it out here and I didn't have anything else to read."

Cecily's mother passed along Willa Cather. "*Oh Pioneers!* and *My Antonia*. She knew I was going out to Nebraska and told me it would be good to read Cather."

These three people violate most of the public's perception of college students. They travel thousands of miles into a completely new environment and culture, work from dawn to dusk at an arcane task that generates a warehouse full of metaphor at worst, homolog at best, for the human condition, then spend their evenings reading Cather and Wilde. And what do they get out of this commitment?

"Most of my assistants have learned what science and research are really all about," offers Charles. "Most of them have taken a lot of courses, but haven't immersed themselves in research as they do out here. And various people have immersed themselves to various degrees, but the notion of the intensity that's required for a successful research project is something they gain. And the hard work. And to some degree the tedium and boredom. But whether they ultimately do social behavior or molecular biology, they come to understand the fundamental principles involved in scientific research. I get a great deal of satisfaction from exposing them to this aspect of the work. Because the majority of students don't learn that sitting in lectures. And a surprising percentage of my assistants have gone on to scientific careers of various sorts."

Do the young people who've taken the leap agree?

"There's only so much you can see in the lab," Cecily says.

"You can also see scientists happy," adds Mike. "The research I've seen back in the lab I hated. But out here research takes on a different meaning, without the four walls around you. It can be enjoyable. It's important for young scientists to know that research can be enjoyable." Yes, I agree. No matter what the walls are made of—money, missions, pressure to perform according to accountants' standards, utility, prestige, political power, fashion—research takes on a different meaning without them.

Part III

Making Their Peace

He was taking a course in plains literature and had brought me one of their texts. I'm glad you're here, I said, now maybe your mom will make some cookies. He ignored the smart-aleck comment and handed me a paperback. I scanned the pages, stopping here and there to read. The weather's always a character, I noted. He nodded. Lots of gloomy cloudy days, lots of wind, lots of snow, lots of thunderstorms and hail tearing up the crops, tornadoes, blistering summers, droughts, you name it, plains literature's got it playing a role. We're the breadbasket, I reminded him, that's why. When you grow annual grasses for a living, then the weather's a player in everything you do. You can't live here unless you make your peace with the weather. You can't live anywhere unless you make your peace with what that place has to offer, he said. He was majoring in philosophy. It's already starting to show, I thought, as I laid the paperback on the coffee table, closed my eyes, and made my peace with a prairie Sunday afternoon while his mom made cookies.

8
Hyalella azteca

Rings on her fingers and bells on her toes
— N U R S E R Y R H Y M E

Once upon a time we went to a roadside ditch. There was *Hyalella azteca*. Then we went to a river. There, too, was *Hyalella azteca*. Then we decided to go to a cattail marsh, where we found *Hyalella azteca*. We traveled many miles to the west, laughing all along the way, wondering whether we'd find *Hyalella azteca*. Sure enough, we did. In the crystal springs, on the algal mats of drying ponds, beneath the bark of soggy rotten limbs, in the river, everywhere, *Hyalella azteca*. What are they good for, we asked, jokingly, because everyone already knew the answer: duck food. *Hyalella azteca* had been put on Earth, shmoolike, so that wild ducks might have tiny shrimp cocktail. Then someone asked what seemed like a stupid question: Why don't we just study *Hyalella azteca*?

The last, seemingly stupid, question turned into a good, no, a great, idea. Why? Because nobody cares, indeed nobody *knows*, except those involved in the deed, if and when *Hyalella azteca* is killed. No rabid animal rights activists descend upon your classroom, overturning microscope tables and pouring gallon jugs of water back into the lake. Nobody cares if *Hyalella azteca* is squashed beneath what to it must feel like an enormous weight, namely a cover glass on a microscope slide. Nobody anesthetizes *Hyalella azteca* before either smashing its guts out by pushing down on the cover glass, or ripping it limb from limb with fine forceps, or even carefully tearing off its head, then teasing out its whole intestine, into a drop of water, to see what *Hyalella azteca* has eaten. You can put *Hyalella azteca* directly into alcohol, formaldehyde, or boiling water; nobody cares. *Hyalella azteca* requires no salaried caretakers, no Institutional Animal Care Committee approval, no inspecting veterinarians. No Federal permit needs to be obtained before collecting *Hyalella azteca* or keeping the species in captivity. No crusaders make secret videos of young people killing *Hyalella azteca*, then release those tapes to the media.

In fact, human beings generally give *Hyalella azteca* less respect than they give to mosquitoes, thus making the study of this species of small freshwater crustacean, along with the organisms that occupy its surface, exceedingly easy. Biologists have to take their organisms apart, if not literally, then figuratively. Even the behaviorist doing fieldwork dissects the lives of birds, for example, cutting their comings and goings into stimuli and responses, substituting causality for beauty, reducing courtship to statistics, the feeding of nestlings to weights and measures, the act of fledging into a measure of Darwinian fitness. But the person who studies symbionts must take his or her subjects apart literally as well as figuratively. *Hyalella azteca* is the most easily obtained, most quickly used, and most numerous of all such subjects. Nor does *Hyalella azteca* complain, scream, or whine when it gets smashed and dismembered. But best of all, *H. azteca* is typically covered with one-celled animals.

The majority of animals on Earth live in and on other animals. Every species that has been studied carefully has been shown to play host to a long list of other species, not only parasitic animals, but also bacteria and viruses. For aquatic species, add diatoms and algae to the list of common if not obligate symbionts. And upon small freshwater crustaceans such as *Hyalella azteca*, the most dramatic of the riders are single-celled, ciliated, ectosymbiotic animals known collectively as "peritrichs." The protozoan order Peritrichida includes some of our most common and familiar microscopic denizens, as well as our most flashy. *Vorticella* may not be as commonly recognized a name as *Paramecium*, but it's nevertheless well known, especially among high school biology teachers and anyone else who's grabbed a handful of grass, put it in a jar, filled the jar with water, and waited two weeks for this biological time bomb to explode.

Vorticella is a common and opportunistic rider; its spring-loaded stalk will attach to almost any substrate, including floating tiny twig fragments, dead insect parts, live insect parts, crustaceans and snails. In the by now reeking two-week-old jar of water and grass, the stalk base may be embedded in bacterial mats so thick they form a gelled scum. In the quiet luxury of this fetid mess, the stalk relaxes, slowly extends; the bell shaped body expands; the ridges pulled down tightly over the mouth spread open; then out come the feeding membranelles—a large circle of fans beating quickly, incessantly, at least until the next mysterious affront sends *Vorticella* into catatonic jerk. The cell contracts more quickly than a human eye can follow, more unexpectedly than the human train of thought can anticipate. Unscheduled, seemingly at random, *Vorticella* jerks, slams shut, snapped back to its base. You wait, staring into the microscope eyepieces. What could have happened, you ask. Was the light too bright? Did I touch the slide? What happened to make that tiny part of nature recoil so rapidly from my interest?

You don't know what happened. You can't discover what happened. You are forever separated from a piece of your world and a flicker of peritrichian history. You could do experiments. You could culture *Vorticella* in large numbers and study their contractile processes, the cellular and molecular events, the fibrils involved, for the rest of your life and still not discover why they reacted so quickly that day you first saw them on a *Hyalella azteca*. Maybe if they didn't have that contractile thread, that myoneme, they wouldn't seem so capable of distancing themselves from your explanations. A lifetime's work on the biochemistry of stalk contraction and you still can't explain the singular event observed once in nature. You can guess at an explanation, make an educated speculation, a conjecture. But there's only one way you can say for sure why *Vorticella* contracts, and that is to set up the conditions under which you make it contract over and over again in response to some stimulus. Then you could impose your cause on the animal, tell people why nature behaves as it does when allowed to behave on its own, based on your observations of how it behaves when you force it to respond to a certain stimulus. A naive listener would think you were being informative. An educated listener would know you were being scientific, and might be right.

Beneath the weight of the cover glass, *Hyalella azteca* struggles, kicking mainly, endlessly kicking, and flexing its body, especially the posterior end. The two pairs of antennae also jerk, up and down, up and down. Your search of the surface moves past *Vorticella* on the back, progressing down toward the sides, where the legs are attached beneath skeletal plates. All you see through the microscope is jerking, jerk, jerk, jerk, pause, jerk, pause, jerk, spiney joints flashing past your circular window, and more riders, specifically members of the genus *Epis-*

tylis. *Epistylis* are attached beneath the plates, right at the base of the legs, the base of the antennae, and around the mouthparts. You have no way of knowing what kinds of currents are produced in nature by *Hyalella*'s kicking, jerking, swimming, and crawling, but intuitively, you suspect *Epistylis* is sort of out of the way of the most violent of these. *Epistylis* superficially resembles *Vorticella*, at least in its inverted bell-shaped body, large spirally-arranged set of feeding membranes, and in the contractions that pull the open end of the bell down into a ball, so that the animal resembles a fat purse with its string pulled tight.

But unlike the solitary *Vorticella*, *Epistylis* is colonial. Somewhere, down in the complex tangle of moving *Hyalella* plates, hairs, spines, and joints, a stalk is attached. Sometime, not too far in the distant past, a single free-swimming *Epistylis* cell selected that point on the surface of *Hyalella azteca* to settle and attach. We have no more of a way of knowing, for sure, why that point was chosen than we do of determining why *Vorticella* contracted. As is the case with contraction, we could discover causes and effects through experimentation, then infer the decision making processes, then reconstruct the histories of single-celled animals. But that history would be ours, built by the information we chose to gather, and not necessarily that of the animals involved. You can construct in the lab the most sophisticated mess in the world, to test the various structural and chemical properties that cause *Epistylis* to settle on whatever you've built. But you can't build a real *Hyalella* in a roadside ditch, and that's where the settlement occurs.

Epistylis, being colonial, also has some properties similar to those of the cliff swallows studied by Charles Brown and his helpers. The fate of an individual cell is tied inextricably to that of the colony. New colonies of *Epistylis* are started by certain cells that break off from the old colony, develop a band of cilia, and swim away. These cells are called telotrochs, or swarmers. A swarmer's decision to alight on one of *Hyalella*'s ectoskeletal plates commits the entire colony that eventually develops from that swarmer to live on that particular plate. An individual *Epistylis* at the end of a branched stalk is only one of many, sometimes dozens, of individuals stuck by that first propagule's decision. Budding of colony members produces new colony members, new stalks, and new feeding currents. The ciliary current of the individual is always influenced by the feeding current of its neighbor, and both are buffeted by whatever currents are produced by *Hyalella* itself, with its swimming, snapping, and jerking.

Epistylis does not occur randomly on the surface, however; most colonies are attached on the ventral surface often near the anterior end, and sometimes beside the mouth. Through high-power lenses we see the water currents produced by *Epistylis* appear like a vortex, a powerful vortex, sweeping in particles, but most mysteriously also sweeping some away. A single-celled animal is selecting particles, by size if not by quality or content. You could easily calculate the volume of water processed by a single *Epistylis* cell; in terms of cell volumes per unit time, the amount of water strained through those frantic membranelles would let us see another side of these animals that live on *Hyalella azteca*, namely their labor, their energy expenditures needed to gather food. But it's the selection process that would suddenly reveal *Epistylis*, *Vorticella*, and others like them, to be infinitely more sophisticated than suspected.

All filter feeders select particles from the streams of water they bring in, typically through the beating of cilia. Most ciliated protozoans—from the familiar almost-household-word, *Paramecium*, to genera with feeding (oral) membranes so large they pull the animal through the water—are particle feeders. Early physiological work on *Paramecium* showed that it selected different particles depending on whether it was well fed or starved. If a single-celled microscopic animal with relatively simple and small oral membranes can adjust its diet according to the quality of its last meal, then the word "hungry" takes on a rather general biological significance. We already use the term in contexts ranging from political to personal, from a summary description of the destitute to a throw-away compliment before digging into an upscale gourmet meal. If a *Paramecium* can get choosy, then maybe we can substitute the socially relevant "hungry" *Paramecium* for the clinically neutral "starved" subject of our experiments. A *Paramecium* capable of getting hungry is quite a different animal than one you toss down the drain when you wash off a slide.

You see, the knowledge about experimental animals only increases one's respect for them. And that is the main point of this story of *Hyalella azteca* and its riders, even though we've taken a short diversion into the habits of *Paramecium* for no other reason than that it has a familiar name. What *Vorticella, Epistylis,* and the others have over *Paramecium* are structural complexity and commitment. The commitment is to a way of life, namely that of the sessile, structurally distorted, filter feeder, the species that, unlike the motile, torpedo-shaped *Paramecium*, remains fixed in place and displays an evolutionary history resulting in a rather radially symmetrical body. The complexity, however, is in the feeding structure. *Vorticella, Epistylis,* in fact all peritrichs, have rows of triangular membranes (membranelles) arranged in a spiral. Each membrane consists of about a hundred cilia. Beneath the membranelles, inside the cell, is a complex network of supporting fibers. Families, genera, and species of peritrichs vary in the details of this feeding apparatus structure, but in each case, the row of membranelles is large, powerful, and impressive. There is no theory to explain the origin of differences between species; the general theory of evolution seems to break down when we try to apply it to sessile, filter feeding, protozoa. And, we've not even begun to decipher the relationships between particle choice and physiological condition in these animals with highly complex ciliary feeding structures.

Ciliary motion also illustrates, perhaps better than any

other natural phenomenon, the principle that small forces operating in large numbers can produce major changes in an organism's environment. Ciliary motion sweeps foreign particles upward out of your bronchial passages (unless you're a heavy smoker). All sponges, bryozoa, phoronids, rotifers, and brachiopods—several entire phyla, some found in both fresh water and in the oceans—feed solely by ciliary motion, sometimes straining bacteria-sized particles from a water column totaling a hundred times their body volume in a day. Most of these groups of animals have been on Earth since the Cambrian period, i.e., for at least 500 million years. The sponges, bryozoa, and brachiopods have magnificent fossil records. The same process that *Vorticella* and *Epistylis* use to feed, has fed an unestimable number of individuals and species that in turn built houses—shells—that later became fossilized. A significant fraction of our vision of the ancient seas rests, in the final analysis, on ciliary motion.

The animals that cover *Hyalella azteca*'s body surface are near the small end of the filter-feeder spectrum. A clam in the same lake as *Hyalella* also strains food particles out of the water, using gills covered with cilia to generate a water current that enters the mantle cavity beneath the shell, follows a highly specific route over the gill surfaces, then exits by way of a siphon extending between the shell valves. In terms of volume, the clam is half a billion times the size of *Epistylis*, yet they eat the same thing for dinner. And in their dining, both accept some particles as tasty, and reject others as unfit for wild animal consumption. Our ability to discover the reasons for acceptance and rejection is equivalent to that of discovering the reasons for stalk contraction and site choice. Yes, the experiments are possible in the lab—feeding particles of various sizes and qualities to *Epistylis* and *Vorticella* of varying degrees of hunger and satiation. From such experiments, we know that size is the main factor in choice. Bacteria are typically one ten-thousandth of an inch long; bacterial-sized items usually get taken in. You can learn how the choice was made in the lab, but you must then assume that's how the choice was made in nature.

Suddenly our study of *Hyalella azteca* and the organisms that occupy its surface become a study in the philosophy of science instead of a study of crustaceans and ciliated protozoans, an exploration of the information that can and cannot be gained from experimentation aimed at the discovery of mechanism. Attachment site selection, contraction, and particle selection are all proximal events carried out repeatedly as a result of ultimate events. That is, they are functions acquired through evolutionary change. Proximal events—functions—are studied using proximal questions that begin with the word "how?" Ultimate events—histories and their consequences—are studied using ultimate questions that begin with the word "why?"

"How" questions demand answers that describe mechanisms, the manners in which events occur. Mechanisms almost always can be deciphered, taken apart, put back together at least somewhat successfully. "Why" questions demand answers that describe deep reasons and causes. Much of modern science assumes that answers to how questions are satisfactory for use in answering why questions. Thus disassembly, reconstruction, and search for mechanism become equivalent to search for underlying causality. We seem to assume that our experiments on stalk contraction in *Vorticella* explain why contractile stalks exist. Our present focus on commercial biology—that intended to solve immediate human problems involving money, political power, health, and food—tends to ignore the ultimate questions altogether, relegating them to the Realm of Unimportance. Yet to a biologist, the ultimate questions are themselves the Realm of Importance; evolution is *the* conceptual framework within which all of biological sciences rest. In the matter of evaluating questions in science, it's a fundamental error to assign importance on the basis of money and power.

But a bigger mistake, I think, is to equate the proximal with the ultimate, the study of function with the study of origin, regardless of how intimately the two might be entwined. That mistake is tantamount to equating method with idea, technique with concept, a city ordinance with history. Such equations defuse science, make it safe and servile. When you teach safe and servile science, you get safe and servile students. When you teach safe and servile science in high school, you get college students who walk out of class at the mention of evolution. Ultimate questions are those of phylogeny, of evolutionary relationships, in which deep causality is sought in the unique geological history of the planet; the boundary conditions under which ciliated protozoans exist, the fundamental nature of resources and the basic equipment a species possesses not by virtue of its need, but by virtue of its ancestry. Ultimate questions, rephrased appropriately, apply to governments and churches as well as to ectocommensal peritrichs.

Among the peritrichs on *Hyalella*, *Lagenophrys* is the one that appears to have chosen security over adventure, in an evolutionary sense. Members of this genus live in houses made of protein. The house looks like a roundish flask or canteen lying flat with the opening turned upward. The opening comes equipped with a two-piece lid that pops open, allowing a large, ciliated, circular feeding device to extend out into the surrounding water. Beneath the microscope, *Lagenophrys* pops open its lid, feeds, snaps back inside, repeating the cycle of feeding and hiding at fairly regular intervals. You can't know whether *Lagenophrys* feeds for long periods with its membranelles extended when *Hyalella* is in a roadside ditch. The hesitancy may be a product of our observation techniques; in order to see the animal, you have to shine a concentrated light beam on it through a lens and compress its *Hyalella* under a cover glass. As with the other peritrichs, indeed as with the smallest of subatomic particles, you're always suspicious that the tools necessary to see an event dictate your perception of that event. Somehow that lesson seems most concrete when you videotape ectocommensal peritrichs, then take the tape home to watch it in the safety of your den. Ensconced on the couch, remote power-box in hand, you can easily switch from *Lagenophrys* popping in and out of its protein house for a quick snack of bacteria, to the latest network news.

In contrast to the other peritrichs, *Lagenophrys* appears to compete effectively for space on an antenna or coxa. But with one-celled animals, there's always a question of whether competitive ability is an evolved function. In the case of *Lagenophrys*, competitive ability comes with the house (called a lorica). These animals could be competing by default, simply because their somewhat flattened circular dwellings occupy surface space so that others cannot. You never see another species attached to *Lagenophrys*. Species of *Lagenophrys* differ in their lorica shapes and modes of attachment. These differences are interpreted as adaptations to occupation of different parts of a host's body. We all need space. *Lagenophrys* tells us that the space we need is determined in part by the species to which we belong, i.e., by our history.

Rhabdostyla has a body that looks very much like both *Vorticella* and *Epistylis*, but lacks a long stalk. Individual *Rhabdostyla* also are usually much smaller than members of the other genera, and end up in odd but strangely interesting places, e.g., right at the joints between leg segments, attached to the tiny membranes between skeletal places, wedged between microscopic spines that on *Rhabdostyla*'s scale look like large spiked fence posts. *Rhabdostyla* is small enough to ride the antenna segments, often by the dozens, occasionally by the hundreds. The distribution of ectocommensal peritrichs on *Hyalella* lends substance to our suspicions about the antennae, of which, like all crustaceans, *Hyalella* has two pairs. In many species, there are obvious structural and functional differences between the first and second pair of antennae, respectively. But in *Hyalella*, the major difference between these pairs is the fact that one is inserted into the anterior portion of the head above the other. At least that appears to be the major difference until you study who lives on these long, fine, jointed, constantly whipping appendages: mainly *Rhabdostyla* and *Lagenophrys.*

The two ectocommensals utilize the resource known as antenna length differently, and *Rhabdostyla* uses the first antennae differently from the way it uses the second. In a bucketful of *Hyalella azteca*, there usually will be over twice as many *Rhabdostyla* on the second antennae than on the first. The protozoan "sees" the second antennae as a more desirable place to live than the first. But having made that choice, do the two antennae differ in other respects? Evidently not, for the distribution along the antennae is virtually identical—most riders locating on the second through the sixth antennal segments away from the head, and the average location of a *Rhabdostyla* being the fourth segment. *Lagenophrys* "sees" no difference between the two antennae, being distributed equally, both in numbers and in location, on the first as on the second. It's an anthropomorphism to suggest that *Lagenophrys'* protective house makes it immune to differences in neighborhoods. But the first and second antenna pairs do differ in terms of the amount of violence they experience. Although that violence is expressed in water currents and whipping movements, *Rhabdostyla* responds to it, *Lagenophrys* ignores it.

Hyalella azteca being so cooperative about being taken apart, a pair of bright students decided one year to remove the first antennae, return *Hyalella* to its jar of ditch water, along with an infected control animal, and just see if removal of the first antennae resulted in a role reversal, so that the second antennae became a *Lagenophrys* and *Rhabdostyla* neighborhood equivalent of the first. The answer was yes, it did make a difference that the "protection" of the first antennae was gone, and no, the students were not convinced they had done the right experiment, or even if they'd done a wrong experiment correctly. They walked away from their project more educated about the subtleties of experimentation, and what can and cannot be learned from it, than about peritrich protozoan crustacean riders. It took them two weeks to do the work, their glassware consisted of recycled mayonnaise and peanut butter jars, and they collected their *Hyalella azteca* while we were out doing some field exercise that had nothing to do with *Hyalella azteca*. That is, they used their wits and exceedingly common materials to learn one of the truly major, and mostly unteachable, lessons of science. No; it was not a stupid idea to study *Hyalella azteca*.

The various riders on *Hyalella* theoretically form a community. They are species that occupy and in some cases, depend on a common resource. Although *Vorticella* may be somewhat of a facultative opportunist, being able to survive on a variety of substrates, *Epistylis* is an obligate ectocommensal, as are *Lagenophrys* and *Rhabdostyla*. That is, insofar as we know, none of these species, except *Vorticella*, lives on nonliving substrates in nature, and from that observation we infer that they cannot *not* live on nonliving substrates. The ectoskeleton of *Hyalella azteca* is a resource upon which these riders are forced, by their evolutionary history, to depend. They go where it goes, when it goes, including down the gullet of a sunfish. Once stuck on *Hyalella*, the riders are a community as surely as those in a commercial jet liner or a wayward bus or a ship at sea.

Community ecology as a scientific discipline rests firmly on the paradigm of interspecific competition. Although that conceptual framework is being dismantled by a few deconstructionists using communities in which forces other than competition seem to be shaping species composition and diversity, the default assumption in ecological circles is that competition provides structure to a community. Car dealers and large appliance stores compete for consumers ready to make major expenditures. Grocery stores and liquor stores compete for consumers who want to get through the next few days and are looking for a variety of means to accomplish that goal. Denominations compete for the religiously inclined; movie houses, video rental outlets, local theater companies, orchestras, and concert series compete for the money you spend on entertainment. Or so we typically conclude when asked about the factors that determine what stores are located where and how long such stores stay open.

But if dollars were suspended food particles, and your hometown swam through an algae-clogged oxbow, shedding its skin periodically, you'd at least ask whether your perceptions about community organizing forces were applicable to the place where you live. Indeed, intuition tells us that cells of an *Epistylis* colony compete more strongly with one another than any of them do, or the colony as a whole does, with other species. It's the salesmen on the showroom floor that are in intense competition among themselves for the customers that come in off the street shopping for new cars. The particles that spend their Sundays looking for a new washer and dryer mean little or nothing to the filter feeders who've selected an attachment site down at the Ford dealership.

How many different kinds of riders do we find on *Hyalella* and how are they arranged on the smorgasbord? In addition to species of *Vorticella, Epistylis, Rhabdostyla,* and *Lagenophrys*—and there may be more than one in each of these genera—we find *Colacium,* a relative of the familiar *Euglena, Vaginicola,* a peritrich that lives in a vase-shaped house, *Carchesium* and *Zoothamnium,* two colonial peritrichs that unlike *Epistylis* have contractile stalks, and *Acineta,* a sessile predator—like a protozoan spider in its web—that waits with poisonous tentacles to catch its prey. *Vorticella* is solitary, and is not an obligate rider on *Hyalella. Epistylis,* although colonial and an obligate rider, does not have a contractile stalk, thus the members of the colony retain their individuality. In *Carchesium,* the individuality is somewhat reduced in that each cell's stalk contracts, and whatever stimulates one stalk to contract will very likely also stimulate those of nearby cells. But in *Zoothamnium* the individual is subordinate to the colony, for the contractile fibrils extend throughout the colony's branching stalk, and whatever fires off one cell's reaction quickly involves the entire group. The various inherited traits, preferred locations, competitive abilities and situations, retention or loss of individuality of all these riders, most of whom are closely related, provide us with almost inexhaustible opportunities to learn how diversity occupies a shared resource.

The computer age strikes us all, even college students determined to have a good time but learn something of their local muddy world in the process. With software comes power, including the power to convert *Hyalella azteca* and its community of riders into a massive exercise in number crunching. Science is, if anything, numerical. If you can't crunch numbers, you can't be a scientist. But the number of numbers involved, and their use in converting the tangible into the abstract, sometimes can be a shock. Twenty people with nothing much else to do can collect a couple of hundred *Hyalella*, then each learns to recognize the genera of ectosymbionts, and counts the numbers of the various genera on each of the crustacean's body segments, each of the antennal segments, each of the appendages, of at least six or seven individual *Hyalella*, all in an afternoon. Then the fun begins.

By midnight, it's become obvious that we've only scratched the surface on *Hyalella azteca*. Does the diversity of parasites vary with age of the host? Is there evidence that any of the pairs of riders competes with one another on one part of the animal but not on another part? Do colonies vary in size according to where their propagule landed? Does one roadside ditch have a more diverse fauna of ectosymbionts on *Hyalella* than the next ditch? If so, what are the qualities that distinguish roadside ditches? Is there any pattern, any defined structure, any order of abundance, of the various peritrichs? Or does it appear as if

Hyalella azteca is playing a multiple-prize lottery in which the winners can get any combination of filter-feeding ciliates on its surface, and in which, unlike in the human games, everyone wins? Can we build a computer model of *Hyalella azteca*, put it in a computer roadside ditch, make it play this lottery over and over again under a variety of conditions, and see what happens? Yes. We can do almost anything with *Hyalella azteca* and its ectosymbiotic protozoans. This little freshwater crustacean, widespread geographically and ecologically, is indeed a teacher's dream.

So what's happened as a result of our asking the seemingly stupid question: Why don't we study *Hyalella azteca*? A lot. Of all the challenges faced by teachers, the most frustrating is that of trying to pass along to your serious students the lessons to be learned from a study of *H. azteca*. Plain and simple, the lessons are:

First, pick something that grows everywhere if you want to make your peace with droughts. Pick something that can be found across vast distances and diverse habitats, and you'll have a kind of organism that gives more than it takes. Such animals do many things for themselves; small wonder they will do much of your work for you. They relieve you of the need for expensive equipment, specially concocted diets, highly regulated temperatures and lights, deionized water, and human helpers. They are weeds in the general sense. All we've ever used to study *H. azteca* is a microscope, a computer, and some miscellaneous recycled glassware. All we've ever asked of our institution was a microscope, a piece of equipment that's been the standard observational tool of biology for three hundred years, and some recycled glassware. *Hyalella* did the rest—had babies in a ditch, a river, a stream, a pond, a well tank, grew into a range of sizes and sexes, but most importantly, acquired a community of riders.

Most of the riders were fastidious, picky about their food and the water currents they'd tolerate and choosy about where, among the plates and spines and hairs and joints they'd settle. The weed *Hyalella azteca* participates in relatively predictable interactions with numerous relatively unweedy species. The second lesson is that with such systems you can study the interactions themselves, that is, ask higher-order questions with some hope of finding a partial answer. What kind of a relationship does *Lagenophrys* have with *Hyalella azteca* is a question of a higher order than: What is this animal? (A: *Lagenophrys*). And the question: What kind of a relationship do *Lagenophrys* and *Rhabdostyla* have with *each other* in the shared environment established by the crustacean, its habits, and its complex movements is of a still higher order than the first. A microscope and a ditch provide the material for the study of questions, and that's what scientists study, ultimately: questions. The higher the order of questions, the more sophisticated the exercise. And this is the intent, is it not, of our enterprise, to produce people who ask higher order questions? Of course. Looking back on that day, "Why don't we just study *Hyalella azteca*?" was of a pretty high order.

9
Worms

There's no god dare wrong a worm.
— RALPH WALDO EMERSON

About 10 million years ago, the Rocky Mountains arose out of the plains and promptly began washing down to the sea, broken into boulders and stones by ice crystals long ago destroyed by a cold sun, worn into sand by winds and rains and thunder, and lightning bolts that flashed and boomed across the young peaks eons before there were people to love thunderstorms; and then carried by the mighty undammed rivers of muddy snowmelt to the oceans, as temporary and changing in their own ways as the morning mist. Along their journey to the sea, rocks became floodplain soil which eventually became silt. Windblown sand and dirt also swirled above the crumbling mountains, occasionally mixed with volcanic ash from enormous explosive holes belching far to the west, and fell into the rivers, later dropping out of still water in the oxbows, congealing into black mud that sat buried and reeking, and seething with worms, before there were people to smell marsh gasses.

But these are not ancient events. Ten million years is no time at all, geologically speaking. The Rockies are children, the latest in a long succession of mountain ranges that have arisen out of the continent's core, above the Stable Interior Craton, then washed away into the prairie rivers. The rivers themselves are, in geological terms, as ephemeral as the rainstorms, but their actions are timeless, made permanent like a birdsong, by repetition: The silt gets picked up, carried a mile downstream, swirled around in an eddy beside an uprooted cottonwood, then dropped. The snowpack eventually melts. In the alpine heights, dwarf willows then lie burning in the unfiltered sun. The river goes down. Algae and slime cover the backwaters clogged with rotten weeds and last year's cockleburrs. The silt turns to mud— sticky, sulfurous, black, rich, putrifactive, wonderful mud. Then the worms appear. Nobody knows where they come from. Two centuries ago they were thought to arise spontaneously from the muck. Today that view is only valid as a metaphor. Now we know that worms have been riding the drifting continents for at least half a billion years.

Back in the lab I stir a bucket with a stick, just to suspend the mud, then pour the mix into a glass one-gallon jar. The blackest mud never pours; you have to scrape it out; you always get some on your jeans; your jeans always smell like that mud, even after you've washed them. Some of the mud always gets under your fingernails; your hands always smell like that mud, too, even after you've washed them. It takes a couple of days for the mud and silt to settle. Tiny snails crawl up the side of the jar; with a hand lens you can see their mouth parts scraping regularly against the glass. Below the mud surface you can see no-see-ums also, and other insect larvae, pulsing and undulating in their little tunnels.

Eventually the mud sorts itself out into a thin brown surface layer, sometimes covered with a light haze of slimy something, and a deeper darker layer that eventually becomes laced with tunnels, a micro-version of the oxbow bottom. An air stone that releases one bubble every ten seconds keeps the water surface agitated enough to prevent the whole jar from becoming a model cesspool. A jar in the lab teaches you more about interactions and stability than any book can; when one air bubble every ten seconds influences a whole gallon jar, then you know that ecological processes are working to provide stability to a system that would otherwise collapse into a putrid mess. If the bubble represents fresh air, then this jar, you think, might also be a metaphor.

You see the worms after the water clears. The larger ones are mostly *Limnodrilus hoffmeisteri*; they emerge from the mud as a small, wavy, forest of rear ends undulating incessantly above their still buried heads. Books claim the rear end waving is a response to oxygen stress; maybe so, but these worms move incessantly regardless of their circumstances, or at least so it seems. Neither *L. hoffmeisteri* nor any of its near relatives can be kept within the field of view in a microscope; their obsessive pulsations carry them maddeningly off to the side, thickening a muscle layer just when you had an interior organ in focus, twisting just at the moment you were ready to make a decision about the tips of their bristles. Sensitive and suspicious beyond belief, they ball up at the slightest change in their environment: microscope lights, pressure on a cover glass, alcohol, formalin, curiosity.

Yet in their writhing, you get glimpses through their body wall of internal tubules, ciliated funnels sweeping in fluids, arteries pulsing with some ancient inherited rhythm, sperm tails a mass of seething motion, tufts of gently curved but forked hairs being swept back and forth by tiny muscles. The microscopic motions completely alter your perception of a worm. No diagram can convey this sense of seeing for the first time, what the world you cannot enter is like. The fact that such worlds occur by the billions, in muck wherever there is muck, only multiplies the feeling you've been excluded by your ignorance, from the soft, translucent, incessantly active, cellular-level universe of primeval life itself. This feeling is the education; diagrams serve mainly to tell you what you miss by studying diagrams.

Why this interest in worms, you ask; leave them be; let them eat mud; nobody makes money off worms. I've come to have no choice but to concern myself with worms; they are entangled not only in the tiny dead plant stems and river muck, but also in the life of an animal I know well. Or so the literature claims. *Myxosoma funduli* is the organism that first dragged me off to the Platte River; it's the spore-forming protozoan two students found when they brought a plains killifish back to Cedar Point, spread open its gill operculum, and asked what those white spots were. That single question was the start of a long and fruitful relationship with a microscopic animal. We toured biology together, hand in hand from the darkness of the electron microscope chamber to the subtle jungles of statistical analysis to a February river, frozen glazed and dangerous. But my many attempts to infect fish on purpose with *Myxosoma funduli* were all failures. A description of what didn't work would fill another book.

Although the life cycle of *Myxosoma funduli* is unknown, as is the life cycle of most of its relatives (collectively known as "myxozoans"), that ignorance does not preclude an examination of the literature, nomenclatural history, and spore structure of the organism. Consequently, *Myxosoma* has been changed to *Myxobolus* by the various powers that periodically change scientific names. When that happens, biologists demonstrate to themselves the real extent to which subtle understandings permeate their business: In casual conversation they use the names interchangeably regardless of what the authorities say.

In addition to the taxonomic revisions, there are also a few reports in scientific journals that suggest the spores must live in worms before the fish can become infected. I have many plans, many good ideas, lying dormant in various files, mental and otherwise, that await the confirmation of this claim. So, after twenty years of studying *M. funduli*, I finally admit that it's time to get off high center, to domesticate the parasites in order to do the studies I know lie waiting. In other words, the time has finally come to test the official version of what the world is really like.

Skip Sterner works as a collections manager in the parasitology laboratory of the University of Nebraska State Museum, a position he has held for about three years. Skip is a careful, responsible person, traits that are not acquired easily, so I can state with a fair amount of certainty that he did not suddenly become careful and responsible upon crossing the Nebraska state line. No, Skip must have been quite careful and responsible long before he arrived on the prairies, even back in the years he, too, began messing around with worms in his own efforts to confirm what the literature claimed. Care and responsibility, dedication, objectivity, and patience, above all patience, are traits required of people who decide to investigate, for one more time, a mystery that nobody seems able to solve. Although scientists are typically a conservative bunch and tend to spend their efforts, as well as their fine qualities, on problems that offer some hope of solution, this involvement of worms in the lives of spore-forming parasites is a published assertion that just screams out for confirmation. Such assertions can make people suspend their disbelief. Needless to say, Skip, being curious as well as careful and responsible, bit.

The parasites of interest make up the group to which my familiar *Myxosoma* (*Myxobolus*) *funduli* belongs, i.e., those living in the tissues, often the gill tissues, of various fish. In the fish, the parasites start from an amoebalike stage which grows into a cyst with thousands of nuclei. The nuclei acquire membranes, thus becoming separated out as complete cells which then aggregate, begin constructing a wall and internal structures such as coiled filaments, and finally differentiate into complete spores. One cyst can produce hundreds, if not thousands, of such spores. These spores consist of the infective amoeba, two "polar capsules" containing the coiled filaments, and two hardened "valves," similar to the shells of a clam, that form the spore covering. Theoretically, upon encountering the proper environment, the internal coiled filaments "fire," becoming entangled in something—the "proper environment"—and the amoeba emerges to begin its life as a parasite. In practice, nobody knows for sure that this set of events actually occurs because so few have claimed, as they say in the business, to have cracked the life cycle. Such general failure, however, is not for lack of trying.

Skip wrote a thesis on these parasites, his project being one in which he tried to use immunological techniques to detect relationships between species of myxozoan parasites that infected a fish called the mottled sculpin in a place called Batise Springs, in the Idaho high country. He reasoned that if two species of parasites were in fact related, genetically, then antibodies made against one of the species should bind to spores of the other species. But if the parasite species were not related, the antibodies would not stick to the other species' spores. He made such antibodies by injecting spores into a rabbit, then "tagged" the rabbit protein antibodies with a fluorescent dye. All he had to do then was mix spores and antibodies under a microscope equipped with a special light and the antibodies would make the spores glow in the dark, provided the two stuck together. By this means, using the fluorescence, he could assess genetic relationships. Skip had an excellent reason to wonder whether the parasite species occupying the mottled sculpin were, in fact, related.

"I didn't describe the one from the ovary," he says. "But I found a new species of *Myxobolus* in the spleen, another *Myxobolus* from the brain, another *Myxobolus* from the ovary, a *My-*

xidium from the gall bladder, another *Myxobolus* from the inside of the gill operculum, and a *Henneguya* from the gills." He also kept sculpins in the lab. "Some of the cysts displaced the brain, but behaviorally, these fish acted just like the others. We had them in fifty-gallon tanks."

Like Tami with her assemblage of friends from damselflies, Skip also had a bunch of undiscovered species to name. Mottled sculpins eat worms, insect larvae, almost anything found in ponds and streams, and they are able to locate worms in the bottom muck by sensing the vibrations made by the incessant burrowing movement, then snapping up the worm with a mouthful of mud. The assertion screaming for confirmation was this: The worms were intermediate hosts for the spore-forming parasites. That is, before a parasite could infect another sculpin, it had to undergo development in a worm. If there ever was an opportunity to "crack" this life cycle, Skip felt he had it at the private pond called Batise Springs. The fish also spend their entire lives within a square meter of the place where they were spawned and raised. Not only was the life cycle completed easily in Batise Springs, it was completed in a highly restricted space.

Surely, Skip reasoned, if the life cycle were completed as claimed in the literature, and if in nature it were completed regularly, easily, by a large diversity of spore-forming parasites all within a square yard of bottom sediment, then I can do it in the lab. So he mixed spores and worms in an aquarium. A year later he admitted defeat. Then he got a job in Nebraska. But he never forgot his failure. The claim screaming for confirmation gnawed away at him, largely because of its audaciousness. These spore-forming parasites, claimed the papers, turned into something so completely foreign and unrecognizable in the worm, that they'd been considered a totally different group of parasites than the ones in the fish.

An analogous claim might be that baby puppies come from robins' eggs. Surely if there were published in the most respected scientific journals, the claim that baby puppies came from robins' eggs, and female dogs gave birth to robins, then some scientist would try to duplicate the claimed life cycle in the lab. In this case, the only difference between the seemingly ludicrous assertion that baby puppies come from robins' eggs, and that mottled sculpin parasites come from another parasite infecting near-microscopic aquatic worms, is our familiarity with puppies and robins. In fact, there is not nearly as much difference between puppies and robins as there is between many groups of invertebrates that look superficially alike. Like Skip Sterner, I have a reason to test this claim made in the literature, namely that baby puppies come from robins' eggs, figuratively speaking. Our years of work on the biology of *Myxosoma* (*Myxobolus*) *funduli* in the plains killifish has come to an impasse; all the problems we can solve without solving the most enigmatic of those remaining, have been solved. All the questions that can be addressed with existing approaches have been answered. But cracking the life cycle would suddenly change the entire picture. Several fine doctoral dissertations await confirmation of this wondrous assertion about puppies from robins' eggs.

I envision myself an old, old man, looking back over his life, thumbing through the black imitation leather–bound theses, their typical bound-thesis smell drifting through the room, bringing back memories of successful students, wet, muddy days out on the river, elegant pictures taken through the electron microscope, simple but equally elegant experiments designed to provide definitive, yes/no answers to insightful questions, answers which in turn suddenly allow the interpretation of massive amounts of data. Yes, those were wonderful days! The only problem is that they haven't occurred yet. I still have to confirm the puppy-from-robin's-egg assertion in order to "domesticate" these parasites, i.e., grow them in the lab and infect fish on purpose, before all those satisfying bound thesis–smelling theses will start to accumulate on my bookshelves. Thus one day I decide, like

Skip Sterner had done, to begin this task. I start the same way he did, by collecting worms. Thousands of worms.

The South Platte River, like all braided prairie rivers, is a highly heterogeneous environment. Within a minute of arriving at the river you can walk through clear rushing water, shallow ripples, chest-deep slow moving pools, drying algae-packed isolated puddles, and a dozen other microhabitats. Each of these habitats has its own fauna, its own mixture of fish and aquatic invertebrate species. Over the course of a year, and over the river's several hundred-mile length, such environmental pockets are created, destroyed, and recreated many times. But in all this apparent turmoil, there is stability: You can always find the heterogeneity, year after year; and, although many small mudholes and oxbows are destroyed, others persist for long periods. There is always a source of worms in the Platte. You can look up and down the river, a mile each way, and see dozens of places similar to the one where you got the mud. Back in the lab, a small dish of this mud is found to contain three or four hundred worms. Extrapolating, you conclude that from any one point on the river, you can see several million, if not several hundred million, worms. On a drive from Denver to Omaha, glancing over at the river periodically, you see several billion worms. There easily could be as many worms in the Platte River between Denver and Omaha as there are human beings on Earth.

The South Platte is not unique in this regard; near-microscopic annelids permeate the fine sediments of freshwater streams, rivers, and lakes throughout the world. Comparatively speaking, the worm fauna of those prairie rivers washing down off the Rockies is relatively impoverished. You'd not find nearly as many species in my cup of mud as you might in an equivalent sample from somewhere else, especially in more southern latitudes. Furthermore, faunal surveys in general show that oligochaete communities are, except for some wide-ranging species, reasonably site-specific in their makeup. About 40 percent of all species are known from only a single locality; over 70 percent are restricted to one geographic region. Less than 5 percent of such worm species are cosmopolitan, or virtually global in distribution; *Limnodrilus hoffmeisteri* is one of the latter. And, although the marine annelids are mostly of a different class from the freshwater ones, coastal sediments are often laced with a mind-boggling diversity of sometimes tiny, but just as often incredibly beautiful, worms.

Freshwater oligochaetes like *Limnodrilus hoffmeisteri* are also sensitive to subtle changes in their environments, including the introduction of chemicals we think of as pollutants. As a result, these worms have been used as "indicator organisms" throughout the world in attempts to assess the effects of various pollutants on freshwater biota. Included in this work, for example, are studies on the accumulations of heavy metals, and alterations in community diversity and species make-up in response to agricultural runoff, industrial wastes, and thermal effluents. In fact, with the possible exception of aquatic larval flies of the family Chironomidae, freshwater oligochaetes have been more heavily used in this regard than any other comparable group of invertebrates. One of the nation's leading worm scientists, Dr. Donald J. Klemm of the University of Cincinnati, states:

> . . . examination of benthic community structure has become a valuable tool for regulatory agencies, water resource managers, and aquatic ecologists in assessing and monitoring water quality and detecting pollution sources.

This value is derived from the facts that 1) many species have established pollution and habitat tolerances, and 2) the worms are relatively long-lived and spend their entire lives as part of the infauna, i.e., denizens of the mud. But I'm neither a water resource manager nor head of a regulatory agency. Thinking back over the years of wadings in the South Platte, I must have detected pollution sources out of ignorance or by accident, not knowing exactly who's been dumping what into the water. "Keep your hands out of your mouth" is as fundamental a rule as "wear old sneakers" when stepping into the river. But I do monitor water quality. High-quality water has plains killifish with parasites embedded in their gills, low-quality water does not. With the decision that it's time to bring the gill tissue–invading *Myxosoma funduli* into the lab, I come to view high-quality water as that with worms as well as infected fish. And, with Skip Sterner's arrival in Nebraska, I have someone to commiserate with on the matter of life cycles involving puppies from robins' eggs.

"We used a variety of worms from the bottom of the springs." Skip remembers his attempted experimental infections well. "These were mainly tubificid worms." Tubificidae is the name of the family to which many of the most common species of freshwater annelids belong. *Limnodrilus hoffmeisteri* is a tubificid; *Tubifex tubifex*, the "nominate species of the nominate genus," is globally distributed and a staple of the fish food industry. The family name—Tubificidae—comes from the mucous tubes that these animals build around themselves. In nature the mucus gets mixed up with dirt particles; a mass of several hundred such worms tangled up around a cockleburr is virtually impossible to separate into its component parts; a writhing, pulsing, Gordian knot that ties and reties itself with every touch of a pair of forceps.

"We used *Tubifex tubifex*," Skip continues, "raised them in captivity, then added spores, right there in the mud with the worms. Left them there about two months, in ten-gallon aquaria with five hundred worms each." I have no trouble reconstructing this endeavor in my mind. "We never found any *Triactinomyxon*."

Triactinomyxon is the puppy that's supposed to come from

a robin's egg. The parasite lives as a cyst attached to the walls of a worm's intestine. Within the cyst, nuclei divide repeatedly, then acquire membranes and aggregate, just like the spore formation process of *Myxobolus* (*Myxosoma*). There the resemblance ends. In *Triactinomyxon*, cells associate to form six or eight spores, packed like beans in one end of a pod. At its other end, the pod is drawn out into three curving spines or long hooks. The cysts burst, sending a shower of these three-spined pods into the water from infected worms. Fish are supposed to get infected either by encountering the triactinomyxons, or by eating the worms. The triactinomyxons (puppies) then turn into myxozoans (robins). The parasites are of veterinary importance to the fish-farming industry all around the world; heavy infections can cause wholesale mortality, especially in very young fish—a million one-inch-long catfish can turn belly up in a week; young trout grow twisted heads and swim in circles.

In theory, if the scientific literature is correct, all you have to do is mix worms and *Myxosoma* spores and wait 105 days, after which time your worm water should be filled with *Triactinomyxon*. Aside from the microscope, worms, mud, water, and spores, the critical items for a test of this prediction is a box of recycled clear plastic jars about two inches high and two inches in diameter. The jars originally contained cultures of various sorts—amoebas, paramecium, small crustacea—purchased from supply houses for use in city classes. After their contents have been used, these little jars with their lids become much-sought-after containers. They're just the right size for isolating snails, collecting algal samples, or doing replicate experiments with culture media, effects of light, temperature, species mixture, and various salt concentrations on microscopic communities, etc. Few experiences illustrate as strongly the essential role that the mind, as opposed to technology, plays in scientific investigation, as the discovery that your most important equipment is a box of recycled plastic jars.

Sometimes it's a shock, too, for the uninitiated to discover that in order to guarantee your supply of paramecium on a certain day, you have to buy them in plastic jars. But such is the nature of whole living animals, even globally distributed, extraordinarily common, whole living animals. There is no way to ensure that you have them when and where you need them except to grow them in large numbers. Even paramecia require tender, loving care. And time. Someone has to feed them and change their water. Paramecia are among the easiest of wild animals to maintain, but they require enough attention so that it's cheaper to buy them than to raise them, unless you have some other reason for keeping the cultures going, e.g., they're your most beloved of all research animals. At the moment, in order to guarantee my supply of *Myxosoma-Myxobolus funduli,* I need the South Platte River. As much as I love the river, I'd like to rid myself of this dependency upon it. That's the problem whose solution produces the doctoral dissertations.

It goes without saying that *Limnodrilus hoffmeisteri* is my first candidate for the intermediate host of *M. funduli. Limnodrilus hoffmeisteri* is a better pet than *Paramecium*; it's a relief to discover that one worm will live for a year in a recycled paramecium jar. A single shelf, stuck away in the corner, holds fifty jars, enough for several replicate experimental infections and controls. Once in a while the worms get a tiny piece of chicken liver to eat. $1.59 worth of chicken liver will keep a thousand *L. hoffmeisteri* alive for a decade. For vitamins and comfort, I add primordial ooze to their diet. Primordial ooze is boiled South Platte River mud. A bucket of such mud, ten dollars' worth of liver, and postage would supply all the nation's college and university biology classes with living microscopic freshwater oligochaetes for a year. But one day, in order to test the puppies-from-robins claim, I stop worrying about the nation's college and university biology classes—other than my own—and start isolating worms from the Platte River. Several hundred worms later, I'm ready to confirm a discovery reported in the literature.

I divide my worm supply into thirds. Safely ensconced in their recycled paramecium jars, two thirds of this supply get fed primordial ooze with a pinch of chicken liver. Half, i.e., one third of the total, also get fed parasite spores. Tens of thousands of spores. On a shelf, I have an experiment that should work like a charm, provided the scientific literature is correct. Those worms that have not been fed spores will be considered the "time *t* control," i.e., they're the ones that you examine at the end of the attempted infection experiment to determine whether the worms you've used were naturally infected all along. The remaining third, the ones that don't get primordial ooze and chicken liver for dinner, get examined immediately. They are the "time zero control," those that will tell you whether the worms you've used were infected in the beginning. At the end of an ideal experiment, neither time zero nor time *t* controls are infected, whereas the experimentals are loaded. I mark a calendar date—November 11—105 days after worm exposure to spores. November 11 is the day my entire realm of scientific research, the direction my reputation and career will evolve in the next few years, and the wise advice I give to trusting young souls who enter my door looking for meaningful work, will all change, if this experimental infection works.

Up to this point, all that's been invested in this project is time, trust, the gas to boil a bucket of mud, $1.59 for chicken liver, and electricity to keep the lights on while I boil mud, search for paramecium jars, and clear off a shelf. I also need something that no amount of money can buy: infected fish, uninfected fish, and uninfected worms. How do you finance your research, is a question I often get asked. With my choice of what to study, is a typical answer. No project better illustrates that answer than this attempt to experimentally infect tubificid worms with *Myxosoma funduli*. If, on November 11, I pull those plastic jars off the shelf and find *Triactinomyxon*, then I'll know that by the following November 11, a major paper will be submitted to an international anonymously peer-reviewed scientific journal. The extra year will be the time it takes to find out whether the *Triactinomyxon* are actually infective to the plains killifish. If the answer is yes, then on the following November 11, there will be another paper submitted, this one addressing the question of whether *Myxosoma funduli* can infect guppies and mosquito fish. But if, on November 11, I find nothing, then I walk away more educated but no less mystified about how a truly common, globally distributed organism completes its life history.

All this discussion of wild worms and plastic jars is beginning to sound like blasphemy. Newspapers and magazines give us the impression that modern science relies heavily on large and expensive equipment, complex chemistry, physics and math, and obscene amounts of money. What the newspapers don't tell us is that scientists define the directions of inquiry, that scientists are greatly influenced by money, that money is greatly influenced by anthropocentrism, and that anthropocentrism is part of a positive feedback loop that includes illness and political power. Nor do the newspapers tell us that the planet itself cares little, in the long run, about money, human illness, and political power. The vast bulk of modern biology occupies itself with an extraordinarily narrow segment of the world at large. There are more species of worms in a cupful of Platte River mud than there are species of organisms studied by all the genetic engineers at a typical large university.

The relationship between *Myxosoma funduli* and *Fundulus zebrinus*, the plains killifish, may be one that both inherited from their ancestors. Killifish as a group are notoriously infected with a magnificent array of parasites—worms that live on gills, body surface, in the ovaries and in the eyes, highly modified crustaceans that clamp onto the gills, single-celled animals of at least three phyla. Although true minnows resemble the plains killifish superficially—they are small streamlined fish—minnows and killifish actually are more closely related to one another ecologically than taxonomically. In the South Platte River several species of minnows share habitat with the killifish, at least during those times of the day and night when scientists go seining. Some of the killifish parasites are extreme generalists; worms that live inside the eyes are supposed to occur in as many as a hundred other species of fish. Other parasites are equally extreme specialists; a microscopic worm is found nowhere else in the river except on the gills of the plains killifish. One truly superb doctoral dissertation awaits my cracking of the *Myxosoma funduli* life cycle. Answer the question: Does *Myxosoma funduli* infect South Platte River fish species other than the plains killifish? Answer that question and you get a Ph.D. Why do you get a Ph.D. for answering such a simple question? Because that question is a key one in the untying of another Gordian knot, namely the intertwined mass of causes, effects, and evolutionary histories that dictate the diversity and dynamics of parasite populations and communities in nature.

A week after I set up the experiment, curiosity conquers rationality and I start checking worms for cysts and the water for *Triactinomyxon*, knowing that the parasites are not supposed to appear for another fourteen weeks. Nothing. On other business, I visit the museum once a week. Periodically, Skip and I talk about myxozoan life cycles, his failed experiments, and the attempted worm infections I have in progress.

"Found anything yet?" becomes his standard greeting.

"No," becomes my standard answer.

Our standard greetings and answers continue through October. Periodically I add a comment that November 11 is the 105th day; we'll give microscopic robins' eggs all the time they need to hatch into puppies before passing judgment on this assertion. On the morning of November 11, I sit down at the microscope beside my collection of plastic jars—time *t* controls and experimentals—and begin looking at water and worms. A week later, Skip and I exchange our standard greeting and answer.

"Find anything?" he asks.

"No."

"Neither did we," he adds, assuring me that I'm not the only one who cannot repeat published experiments.

This seemingly unimportant, and unsuccessful, little infection experiment illustrates one of the major factors that dictate the directions and contributions of science in general and biology in particular. The material itself often determines what can and cannot be done with it, and for that reason, biologists tend to pick materials that allow them to explore certain phenomena, that is, to do what they want to do. Then they claim the results of their explorations are widely applicable. Given enough materials and enough scientists, widely applicable discoveries *are* eventually made, if not by individuals on purpose, then by the whole scientific enterprise. This is one of the major reasons why societies that scrimp on science and demand instant results inevitably fail in technology-based economic wars. This is one of the major reasons why politicians who ridicule a research project they can't understand are dangerous to such societies. These are strong statements to emerge unannounced in the middle of a discussion about worms. But they are exactly the thoughts that boil away in your mind out on the river when you're learning firsthand what biology is all about.

Nearly twenty years ago, when two students showed me an infected killifish, it was obvious that they'd discovered the materials to determine how environmental change influences the transmission of infectious agents. The Platte River's middle name is fluctuation. Streamflow varies over an order of magnitude, plugging along at less than 200 cubic feet per second some years, and at more than 2500 cubic feet per second in other years, near our collection sites. Furthermore, such differences can occur over the space of weeks or months; alternatively, high- and low-water years can turn into high- and low-water decades. Both fish and parasites survive this constant change occurring over several scales of space and time. The genetic information contained in the parasites is transmitted by mechanisms defined by the information itself, through this ever-fluctuating matrix of water, mud, worms, fish, and algae. The parasites of the plains killifish provide a model for the dynamic flow of information in all fluctuating environments: rumors in times of war, innovation in times of economic depression, utopian schemes in times of political unrest.

After twenty years of work, the studies show that the probability of infection varies inversely with the amount of water flowing in the Platte—when the water goes up, transmission rates go down, and vice versa. Evidently you have to study a process like that for a long enough period through both good and bad times, plentiful times and droughts, in order to discover how organisms make their peace with an ever-changing world. But eventually you'd like to elucidate the mechanisms by which rumors, innovations, utopian schemes, and myxozoans move through populations of susceptible hosts and in order to do that, you have to be able to complete the life cycle on purpose, on your own terms. So far we've failed. I don't know exactly how the plains killifish gets infected with *Myxosoma funduli*. I know what others *tell* me about how that happens. And I know how Skip and I got infected with the idea that we could do this experiment, namely by listening to others tell us how our little system worked.

10
Droughts

Dreams in Dry Places
— TED KOOSER

We never really were able to reconstruct the set of events leading up to Bill's decision to spend his life with leeches, but all of us—the people who worked with Bill, his parents, his teachers, especially the teachers—spent a great deal of time thinking back to the months before he became interested in leeches, wondering what happened that turned him on, trying to remember something he'd said, some subtle hint, a particular week when the expression on his face changed from that of a typical college student to that of a dedicated, indeed driven, professional scientist. Of course he still hadn't acquired all of the necessary tools—the statistics, the frustration of failure and rejection, the thrill of his first accepted manuscript, the electron microscope techniques, and so forth. But in his mind he was a college student one day, and a professional scientist the next. We tried to joke about Bill's transformation: If you could discover a process by which Bills might be manufactured, such a process could be patented. Usually the joke didn't get many laughs. We all knew it was no joke.

Bill came on the heels of another leech lover, an older college student—Gale—who had many wonderful tales to tell about his past, and who also, for some unexplained reason, decided to study leeches. Gale could hold you enthralled for hours about his experiences with the military, the subsequent experiences with the Federal bureaucracy, his time spent at various life roles that none of us thought truly existed. We offered up a thousand explanations why he was so fascinated with leeches; we also spent a lot of time talking about the classic film, *The African Queen*, and the study of animals other people thought were dangerous. Gale's leeches were covered with babies, potentially dangerous babies, so we talked about science-fiction movies based on dangerous offspring, too. But only one reason for Gale's fascination with leeches stood out in my mind: He'd discovered a way to make leeches stay flattened without the use of drugs. Having made the discovery, in fact a technological advance, then it became possible for him to ask certain questions with the hope of actually finding a satisfactory answer.

Bill, of course, preferred drugs—for leeches—zonking them with barbiturates. But that technique came later, some time after the snuff and cigarette users gave up on nicotine—for leeches—and well in advance of the truly modern era of Cedar Point students, whose preferred annelid narcotic was diet pop. Before Bill started using drugs—for leeches—Gale used hot water and physical force. You see, he said, a leech is just like a person. You know how you relax in a hot bath? Well, a leech will do the same thing. At least that's my theory. Just put 'em in warm water until they relax, then smash 'em between two petri dishes. He demonstrated, using the larger petri lid to hold the warm water and leech, and the smaller bottom turned upside down, to smash the leech flat. You have to hold 'em together, then pour out the water, then pour in some formaldehyde. Then you have to hold the whole thing very tightly together until the leech dies. Then— he demonstrated—you take out the dead leech which still won't stay flat and put it between a couple of slides, then wrap a big rubber band around the slides real tight, and put it back into formaldehyde until it stays flat. He finished with a smile.

The smile was not one of self-satisfaction, but the same kind of smile that must have passed across every scientist's face throughout all history at the instant of technological breakthrough. *There*! said the smile; *that's* how you do it! The smile hid the disdain, stimulated by some experiences in this person's deep past, for those who couldn't figure out something that seemed so simple in retrospect. We excused the disdain. Big breakthroughs always seem simple in retrospect.

If there are professional leech biologists reading this book, they are cringing at the description of this "technical advance." Nevertheless, for Gale, the technique allowed him to eventually see what he wanted to see, namely parasites inside leeches. Suddenly a project he had in mind became possible. So Gale the student enlisted the help of another student, Liz, who eventually became Marv's wife, rendering her insightful judgment that afternoon in a local bar, on the problems grown-up boys have finding play that looks and pays like work. But at the time she was more interested in a good class project than in Marv, and big technological breakthroughs suddenly made Gale's project on leech parasites look pretty good.

Now, several years after all this story took place, when I go back through the hundreds of leech slides made that summer, the vast majority of them have Liz's initials—CEL—scratched into the glass, along with the year—1988—and specimen numbers. The year is a special one, a year of parasitological abundance before Martin Bay Pond dried up, a year in which I made decisions and almost carried them out, e.g., suggesting to a student that Martin Bay Pond might be a great place to begin work on a several-year doctoral dissertation. Now Martin Bay Pond is gone and instead of a place where I can take people like Bill, and Gale, and Liz, all I have are the leech slides. But the slides are exceedingly important; they now form a significant part of the teaching collections for freshman biology classes at my university. I try to tell people what CEL–1988 actually means, but never feel as if they understand the significance of those three letters and four numbers. A student did something that provides generations of students that follow with a window to the world. The window may be small—the proboscis of a leech, or an intestinal diverticulum, a group of eyes arranged in some taxonomically important way—but that window is open. I remember we started calling Liz, "The Leech Queen."

The name was an admirable, traditional, and honorific one, passed down from a couple of still earlier students—Midge and Mary Ann—who had also been known as the "Leech Queens." Midge and Mary Ann had started their relationship with leeches by butchering a very large snapping turtle. Covered with turtle blood, their hands filled with entrails, one of them noticed a leech clinging to the skin, up in the folds between the hind leg and body. Of course they saved the leech. Later, studying the turtle blood beneath a microscope, they discovered that the animal had been infected with an intracellular parasite of the genus *Haemogregarina*. These parasites live inside turtle red blood cells, the intracellular stages being gametocytes that transform into gametes in the intestine of a leech that sucks turtle blood. Of course Midge and Mary Ann would settle for nothing less than an experiment to find out what happened to these parasites once they were ingested by the leeches. Thus began a search for material. The turtles were somewhat of a problem, but just by being alert and brave, the Leech Queens managed to salvage a few snappers from the bottoms of various sulfurous muckholes. The leeches were no problem whatsoever; like anyone who wanted them, Midge and Mary Ann went to Martin Bay Pond.

Had I given much thought at the time to the ease with which Midge, Mary Ann, Gale, Liz, and finally the truly serious leech scholar Bill, became hooked—intellectually—by these rather remarkable predators and blood suckers, then maybe I'd have looked upon Martin Bay Pond as something more than the most bountiful teaching laboratory on Earth. Instead, it was simply a guaranteed source of leeches, mainly *Placobdella parasitica*, as well as a few other genera and species—*Erpobdella punctata* with its elegant undulating swimming motion and apparent searching and decision-making power, and the giant *Macrobdella decora*, dark chocolate brown with bright orange spots. "We found the most beautiful leech!" they'd say, weeks after gathering around the pan with *Macrobdella decora* stretching and contracting over a several inch range. At the time I thought Martin Bay Pond was just a permanent and predictable source of captivating material, sort of like a pond you'd find at any field station, not subject to various budgetary whims—natural and otherwise. There would always be a Martin Bay Pond, I thought, just like there would always be a McGinley Pond, an Ackley Valley South marsh, and a long list of other wondrous and easily accessible muckholes. But like all prairie thoughts about water, these were quite naive.

Tom McGinley seems like a natural source of wisdom about possible leech sites, having been, like most ranchers, intimately acquainted with various isolated bodies of water. Nowadays he lives in a house behind the store where he sells fishing licenses, park permits, pop, groceries, RV and trailer hookups, laundry privileges and showers. Although the operation is owned by his son, the store sits on a small piece of land that must have been McGinley Ranch property before the power district and the state took over the North Platte River valley and turned a stretch of it into Lake McConaughy, Lake Ogallala, Lake Keystone, the Sutherland irrigation canal, and the fee camping area. Lake McConaughy is twenty-two miles long; Lake Ogallala is a borrow pit for the dam thus is fairly deep and long, paralleling the foot of Kingsley Dam. The McGinley store is located along a curve in the Keystone Road, near the north end of Lake Ogallala. Except for the blizzards and gulley-washing hailstorms, the setting is a peaceful and idyllic one. Behind the store, Kingsley Dam rises like a towering battle over water rights and relicensing applications. In comparison to what the land was like when Tom McGinley's grandfather homesteaded it, this section of the North Platte River valley is fairly occupied, although deer still wander through the campsites and coyotes serenade the city folks out for a day or two of mosquitoes and sun.

Tom McGinley has adjusted his business operations considerably in response to a variety of events not atypical of those experienced by large family-owned ranching operations. Martin Bay Pond, McGinley Pond, Whitetail Creek, and any number of other wet spots, including a well tank full of salamanders, are all part of the original McGinley Ranch property. For thirty years, Tom himself lived on the Buckhorn Springs Ranch, which forms the western boundary of the Sillason property. The Buckhorn Springs buildings are nestled in a cottonwood grove north of the road to Nevens; this section of what I sometimes call "Laura's road to the heath" is narrow and overhung by willows, a tortuous trail made mainly of blind corners. The ditch willows are a dead giveaway—somewhere just below the surface is water.

"We had a corral just east of the barn. We used that corral for branding and vaccinating." Tom reminisces about his seasons as a rancher. "That underground spring was just east of the corral. But they move. They move around under the ground. This one started moving toward the corral. Before long, the cattle were standing knee deep in the mud. Then it kept moving toward the barn."

The Buckhorn Springs barn has a concrete foundation about eight feet high and two feet thick.

"You know how they have to put those drain tiles under the fields back in Illinois? Well, I thought we could put some of those in under that corral. We started digging a trench, but it always filled up with water. Eventually we had to make a trench fifteen feet wide in order to get three feet down. Then we laid some of that four-inch perforated pipe in the bottom. We put a 'Y' in about every ten feet and drained it back out to the east."

"Did it work?"

"Sure. But then we dug under the foundation of the barn and laid another pipe under the foundation."

In a land called "the Great American Desert" by the nineteenth-century settlers, a twentieth-century rancher has to put in a drain pipe in order to save his corral and barn. Tom's grandfather, George McGinley, homesteaded the ranch in 1888, the year of the Great Blizzard celebrated in Nebraska folklore, poetry, and abstract mosaic art in the state capitol. Tom doesn't remember his grandfather's talking about the Great Blizzard, but his memory of leech heaven, Martin Bay Pond, adds a certain philosophical perspective to mine.

"Before the dam was built, and that was started in 1937, the railroad wasn't there. To the best of my knowledge, I don't recall there being any pond there, but then I didn't live down here. Then when the dam was put in, they moved the railroad out of the valley and up on the high ground there. And as the dam filled, and got clear full—I think they declared it pretty much full by '48—that pond developed. We had a lot of wet weather in the middle and late forties and the pond that developed there was maybe ten foot deep.

"Then in the middle fifties, starting about '52, we started to have a drought that covered the western half of Nebraska, and went clear to the continental divide. So the North Platte River basin didn't shed much water. They were letting about the same amount out down river for irrigation, so the lake went down, and that pond went dry. The lake went down to about sixty foot of water in the morning glory; that was in 1955, '56. The pond went completely dry. In fact my uncle, George McGinley, was operating the ranch at that time and had planted cane on it for cane hay. Year was '55 to '56. Late '56 and early '57, we started getting a lot of rain and a lot more water started coming down the river. And in '57 he planted some more cane in there but he never got it harvested because the water started coming up. And then by '58 or '59, the pond was back almost full. The dam was almost full again, too." The local chamber of commerce refers to gigantic Lake McConaughy as Big Mac; Tom calls it "the dam;" the "morning glory" is the outlet tower.

The hydrology of the Sandhills, as well as that of the adjoining Platte River Valley, is a complex subject about which almost everyone has an opinion but about which a relative few have detailed knowledge. One result of this gap is the growing list of rural community water supplies contaminated with nitrates from local agricultural operations. The surface to ground water connections are not always as apparent as they were to Tom McGinley when he lived on Buckhorn Springs Ranch. But when an entire village, isolated far into the idyllic heartland, has to start buying bottled drinking water then you know those connections exist.

When they sink a well, High Plains irrigators are mining a limited fossil resource. Above the Stable Interior Craton of pre-Cambrian rocks lies a series of mud and gravel formations laid down first when most of the continental interior was covered with relatively shallow seas, and later when early- and mid-Cenozoic violence dumped cubic miles of volcanic ash and giant meteorite debris across much of the region. The main water—bearing sands began washing down off the mountains about 17 million years ago, eventually accumulating as an underground alluvial plain that stretches in a two-hundred-mile-wide band from

Odessa, Texas, to Valentine, Nebraska. Volcanoes, meandering streams and rivers, wandering underground seeps and springs, and eastern glaciers that must have blocked these rivers, producing lakes a million times the size of Big Mac, all made their marks on the subterranean physiography.

These marks are still fresh. In their professional modes, petroleum geologists live back beyond the Permian; but High Plains hydrologists speaking of Sandhills groundwater talk in terms of thousands of years. Fossil shelled amoebas from hot Mississippian coastal sediments capture an oil man's fancy; endless eons of stable simmering algal growth and subsequent gentle buried incubation form his picture of an ideal world. But inland water people conjure up woolly mammoths and suckling rhinoceri trapped by sky-blackening Vesuvian ash clouds; buried prairie water is ephemeral, a product of turmoil, a "resource" more akin to cloud patterns than to coal and oil. Those who daily walk the sandy surface understand this fact better than those who sit in offices and make economic plans. When a wandering underground spring threatens your barn, and your pond goes dry long enough to plant cane but doesn't stay dry long enough to harvest it, then you get a firsthand definition of this term "resource." And when you can no longer tell the Bills, Gales, and Leech Queens of the world just to "go over to Martin Bay Pond," then that firsthand definition—of intellectual resource—also burns itself into your brain.

Tom McGinley's seen both sides of the water picture. When he wasn't trying to get it out of his barn, he was trying to get it into his cattle tanks.

"While I was there on Buckhorn Springs, I was going to drill a well north of Keystone. They'd drilled on this piece of ground for test wells—Haggard Drilling Company—thirteen or fourteen wells in that quarter section. They'd go down about forty to fifty foot, hit Brule clay, and get a three-hundred to four-hundred-gallon well. But the well man was visiting with me, and said, 'why don't you go down and witch those places and maybe you'll find an underground spring.' And he pointed out there were two good wells on some other property nearby. So I said, 'What do you mean witch it?' and he said, 'I'll show you.' And he took these two rods, they were just quarter-inch steel rods, about like a long bolt but smooth, bent in a right angle. One arm was about eighteen inches long, then the place where you hold 'em was bent at a ninety-degree angle. You'd hold these in your hands with the short end in a vertical position, and with the long arms in a horizontal position. Then you'd walk along, and when those two rods crossed, then you were over an underground water system."

"What would have been the original manufacturer's use for those rods?" If you could buy equipment to find springfed ponds, even underground ponds, I reasoned, then you'd have a never-ending supply of places to send Leech Queens and Bills. And if these ponds moved around as claimed, and if you knew where they were and watched for them, maybe they'd come to the surface periodically, so that when one opportunity dried up, maybe another one would appear.

"Just a piece of rod, just mild steel, cold-rolled rods. Buy those any place in town." Sometimes I forget that Ogallala, Nebraska, population 5,600, eight miles south of Tom McGinley's store, is a ranching community as well as a tourist attraction. When you can buy quarter-inch mild steel rods in a town the size of Ogallala, that means someone has a regular need to make his own machinery parts. "But you don't need rods that good. If you want to make a rod cheap, just use any piece of wire; coat hangers'll work just the same. Just a piece of metal. Hold 'em very loosely; just make sure they're level. And when you go over the water, they cross. Then back off, and walk back over it at a forty-five degree angle and hit that vein, both rods will go to your right. Turn around and go back, they both go the other way. And he showed me how to do this over a known creek. I don't remember the man's name—Vince something—worked for Haggard; he was their head well driller. They'd drilled thirteen wells. So I went down on this quarter section and walked around for awhile and found a place where these rods crossed. He came out and drilled a test well, went one hundred fifty foot into some good rock. So then he drilled an irrigation well one hundred and sixty foot, with about sixty foot of perforated pipe, and test pumped four thousand gallons a minute."

I wondered if he'd found a hole in the Brule clay.

"It's a place in the clay where there was an underground creek running through. You'd go down in the other places and hit clay and that was it. But when they dug into this underground creek, they started getting some sand and gravel, and then further on down they hit some good rocks. By the time they got to a hundred feet, they were getting rocks four inches in diameter, and it was a really, really good well. So I went down to where a neighbor had a good well, and another good one half a mile west. So you could look at those and see that all three were almost in a straight line."

As I listen to this story, a picture of that creek on the surface begins to form in my mind. There's a wall of thunderstorms in the west; lightning hits along the dark horizon, and a minute later comes the rumble. Small fish dart upstream as the mammoth herd walks into the water, their massive feet crushing the gravel, pushing it out into basins that quickly fill again. The fish are some of the same species I study today. They have worms on their gills, the same species of worms that my students study today. Down in a deeper channel are the four-inch rocks, covered with caddis fly larval cases. A small herd of horses follows behind the mammoths. The rocks are the same mixture of granite and quartz that you find now in surface streams; the caddis flies are probably the same species, too. The wind picks up. The mammoths turn away from the stinging dust and sand. You never know when you choose a four-inch rock from a Sandhills stream, whether that same rock got stepped on by a woolly mammoth. That's why I collect particularly interesting stream rocks. Some are used as paperweights back in my city office. Others just sit on the window sill. Someday, if history tells us the truth, that window sill and my desk will be gone the way of the mammoths and that stone may once again be one hundred and sixty feet below some prairie. Such thoughts build up, like thunderheads, when people like Tom McGinley talk about witching water.

"So I went back to satisfy my own curiosity," he continues, "and went to the northwest, and about every half or three quarters of a mile I found out where this water system went. It went all up into the Sandhills. Then just for fun, my wife and I were running around in our pickup looking for these systems. She would sit in the pickup and hold those rods. And when we'd go over those systems, the rods would cross. Then after that, whenever I wanted to put in a new stock well, we'd go out on the ranch and drive back and forth in the pickup and find where those sticks crossed. And every one of those wells were good ones. I never had a poor producer."

"Did Vince do this for Haggard?" It would be in the vested interests of a well drilling company for its chief driller also to be its chief water witch.

"Oh yeah, but when someone made fun of him, he'd just pass it off. Not too many people doubted him on it. He drilled a lot of good wells. But you can find all kinds of stuff with these sticks. You can find a water line, or an electric line buried especially in a casing. On a gravel road out there by Buckhorn Springs there were some old culverts that were abandoned. And a guy told me those culverts were there. They were old narrow ones that just got buried when they widened the road. So I wanted some culverts for the ranch. The county man told me I could have them if I could find them. So I took the witchin' sticks and found 'em. The sticks would cross. I found three of them. Took a back hoe and dug 'em out."

"Yes," he answers my question about Martin Bay Pond, McGinley Pond, and a little marsh by the Keystone Road, with a generalization, "I believe that coming out of the Sandhills there's a certain number of underground water systems, lakes or creeks, that are even underneath our major creeks. I believe there is a vein o' water up there that starts—I don't know how far back in the Sandhills it starts—but it starts to get close to the surface as it gets close to the North Platte Valley. At different places for different reasons it pushes out of the ground and causes a spring. And that same underground river or creek, it's my belief that it drains into Lake Ogallala, because the northwest corner very seldom freezes over. And that water is warmer and it keeps that end thawed out. All the years I've been here, that end of the lake has never frozen."

From the northwest corner of Lake Ogallala, a depression runs northward into the Sandhills, crossing the blacktop at the bottom of the artificial hill made by Kingsley Dam. Nobody worries too much, or too seriously, about Kingsley Dam. But there's a lot of disaster talk mixed in with other rowdy conversation whenever young people start competing with one another for attention, maybe on the morning ride out to Ash Hollow, or to one of the sandy bays along the north side of the lake. What if the dam breaks, they ask. What would you do? That's a good question, I answer; depends on where you are. These questions usually surface a day or two after we've used the Toe Drains for a class exercise of some kind. Toe Drains drain, of course, the foot of Kingsley Dam, carrying seepage, like Tom McGinley's Buckhorn Springs corral-savers, into a buried pipe then discharging the water at four or five places along the eastern downstream base of the man-made earthen mountain. Toe Drain water is exceedingly cold, coming as it does, from the bottom of Lake McConaughy. The fact that it comes from *beneath* the largest earth-filled dam in the world is not lost on anyone.

Northeast of the place where Tom's underground system

breaks through the land surface and falls into Lake Ogallala, up along the depression and wet trace that points to the dry flats we still call Martin Bay Pond, lies a still more remarkable hydrological phenomenon.

"Tell me about this closer one," I ask, "the little shallow one that had the duck blind on it."

"Here's something that's kind of strange about this whole thing. The little one on this end is fed by springs that come out of the ground just above it, approximately a quarter of a mile above it. This little pond had a duck blind on it from when I first moved to Keith County to live, which was in 1955. During the time when the big lake was down to sixty foot of water, that little pond had water in it all the time. It's a lot higher than the level of the lake. Close to forty-foot elevation higher. I have a feeling this little water system has somewhat of a fall in it and it probably falls rather rapidly down to the river. But nevertheless, during the drought of the '50s, when the lake went almost dry, that pond had water in it. And then all through the years when we had good water levels in the lake, that pond was always there—until this last drought in the late '80s. Now that pond's had no water in it for a year. But now that the lake's coming back up, those springs are starting to seep a little bit. Now maybe fifteen to twenty head of cows can stay alive on the water that's coming up."

"Drinking out of their own footprints." I know those footprints well; they are perfect mosquito breeding places.

"Yeah, just sipping out of the ground." Tom pauses; I imagine him imagining a cow sipping out of the spring water welling up in its own hoofprint. "So I don't know what's going to happen to the ponds if the lake fills up. It'll be interesting to see if the lake fills up whether all these little ponds will come back completely."

Yes, I think, it will be very interesting to see if they fill up. If they do, we'll have a very good opportunity to see how those ponds reestablish themselves. If they don't, then the emptying and filling of the big lake is not a true barometer of good and bad times after all, at least not if you're a bug biologist.

Tom's daughter comes to the front door with a blue official-looking folder, asking whether he has any more daily park permits in the store. He says no, and goes into the house for another booklet of blank permit forms. When he re returns, he has his "witching sticks"—two perfectly bent, well-handled steel rods.

"Here. Noticed these inside. I'll give you a demonstration." Around to the north side of the building a sewer line runs between the trailer park and the women's showers. He holds the rods perfectly level, the short parts resting in his fingers, his thumbs held back as if to cock the hammer of a pistol. He walks across the sewer line. The rods swing sharply toward one another, becoming parallel, pointing in opposite directions. He hands me the pieces of metal.

The long arms are pointed straight ahead, parallel, aimed directly to the west. As I walk across the sewer line, they quickly rotate toward one another, the rod in my right hand pointing south, the one in my left hand pointing north. Tom smiles. Beyond the sewer line, I reset the witching sticks and walk east; again the rods cross. I am completely convinced that I've done nothing to make this crossing happen, nothing, that is, but walk across Tom's yard where he claims there's a buried sewer line. I hand him his witching sticks. He smiles, leans back against the wash house wall. I wish for my camera. Tom McGinley against the white stucco, in his overalls, yellow mesh cap, sunglasses, rods in his left hand. The photograph is entitled: Tom McGinley at home with the weapons he uses to defeat droughts. I wish I had some of those rods. If I had some witching sticks, maybe I could find some more places to send the Bills, Gales, and Leech Queens of the world when they come into my office looking for something to do.

"There's more water flowing into the river from underneath the ground than from on top," he says.

But if you're a leech, I think, you need more than a hoof print full of mosquito larvae. If you're a leech, you need a Martin Bay Pond. Back at Cedar Point, I go digging around in the old garage for survey flags, "ecology flowers," we call them, those springy wires with bright plastic squares that seem to seed themselves wherever some project on plant population biology is set up on the prairie. Selecting two perfect ones, I snip off the flags, measure up exactly six inches from one end, and at that point bend the rods at a right angle. Some day, I think, there'll be a need to find a muckhole. If Tom McGinley can walk out across his pasture with a couple of bent rods and find places to keep his way of life alive, then I can, too, walk across campus with my now deflagged and bent ecology flowers turned into witching sticks and find a perfect spot for the next leech lover that comes along. Then we'll discover if there's a good well to be drilled under the stadium.

It was in the fall when Bill returned to campus. He'd gone to Canada to gain the expertise to guide his study of leeches, used his own kind of witching sticks to locate a place he could sip out of the ground, so to speak. The place turned out to be Algonquin Park in Ontario, and he found water in places with names like Kioshkokwi Lake, Three-mile Lake, Lake of Two Rivers, Marsh by Opeongo Road, Lake Sasajewan, and the Oxtongue River. Four years after he made the decision that he would ride a leech into his future, he hands me a book. The binding is that familiar imitation leather. The pages are that familiar 8½" × 11" high-quality bond. The whiff of odor as I open it is that of the familiar binding and photographic mounting cement, the rag bond, and the ink. This book is entitled *Morphological, Histochemical, and Ultrastructural Characterization of the Salivary Glands and Proboscis of* Placobdella ornata, Placobdella parasitica, and Dessorobdella picta (*Rhynchobdellida: Glossiphonildae*). The author is William Moser. Bill goes to Wordsworth for his epigraph:

> *My question eagerly did I renew,*
> *"How is it that you live, and what is it that you do?"*
> *He with a smile did then his words repeat:*
> *And said that, gathering leeches, far and wide*
> *He travelled; stirring thus about his feet*
> *The waters of the pools where they abide.*
> *"Once I could meet with them on every side;*
> *But they have dwindled long by slow decay;*
> *Yet still I persevere, and find them where I may."*

The slow death of expertise in basic leech biology and natural history is little different from that found when one goes looking for experts to classify gastrotrichs, rotifers, nemertenes, acanthocephalans, or any of dozens of other seemingly obscure groups of plants and animals. The people who know how to do systematics on the world's fauna and flora are dying off. This loss of knowing power is in sharp contrast to our need for knowing power, as studies of tropical forests multiply, almost daily, our estimates of the number of undiscovered and undescribed species on the planet. Such loss is also contrasted with the highly contemporary use of leeches as sources of anticoagulent and as subjects of neurophysiological research. For this reason, *Hirudo medicinalis*, the medicinal leech, has shown remarkable staying power in medical history. In its salivary secretions are found substances that prevent clots. Blood clots are among the complications leading to any number of health problems from stroke to heart attacks. Did Bill decide to save humanity from stroke complications when he started on leeches? No. He just loved worms.

Leeches can be somewhat selective about their prey and

blood sources, some feeding only on fish, others predominantly on snails, and still others on mammals. Until Bill started on his thesis, there had been no major study of the salivary glands of leeches that fed mainly on amphibians and reptiles. The glands are actually single cells, each salivary cell having a body and a ductule. The leeches he studied have a proboscis that pierces the skin of a victim. The salivary ductules form a bundle that enters the proboscis near its base.

In order to get to these glands, Bill had to do some rather remarkable and fine dissections. His drawings are elegant renditions of dissections that are themselves rather elegant. Beside the drawings are photographs—proof that he'd actually done the work—but his interpretive drawings would have been acceptable in turn-of-the-century German journals. Biology is losing people who can make those kinds of drawings as rapidly as it's losing people who know or care what the drawings are. I turn pages until I get to the Scanning Electron Microscope (SEM) pictures. The SEM produces a shock of familiarity coupled with an affront to your ignorance. The structures are familiar—a proboscis looks like a proboscis; but the surface details revealed by SEM are always the shocking part. The cellular coverings and protein sheaths of animal organs are always more ridged, convoluted, pitted, papillated, fuzzy, folded, and wormy than ever imagined—reminiscent of Gulliver's description of giants' skin, and of Gulliver's skin described by the Lilliputians. In both cases, the magnifying glass stripped away pretensions of smoothness. Swift's Gulliver may have discovered shattered perceptions on his journeys, but Bill found the openings of the salivary ductules on the tips of his leech proboscides. His SEMs show an *en face* view of proboscis tip, looking into that dark hole gullet. If you're a turtle red blood cell, that's where you'll go to your death.

The transmission electron microscope (TEM) converts living tissue into abstract art, Jackson Pollocks done in shades of black and gray. These pictures are the micrographs (TEMs) found throughout biology textbooks and learned journals alike, and they reveal what is called ultrastructure—the finest, most minute, cellular architecture observable by existing technology. Ultrastructural views of "the cell" (typically from a rat's liver, thus not necessarily a typical cell) have become biological icons; you don't know your catechism if you can't draw a mitochondrion and rough endoplasmic reticulum. Bill's TEMs of leech proboscis thus showed, as expected, the more arcane view of cells—the view to which a professional biologist can relate, more distanced from that of the familiar SEMs. In the TEMs we see the various granule types, deep invaginations of epithelial linings, mito-chondria, folded cell membranes, cellular junctions, ultrathin slices of secretory granule-packed salivary cells looking like flagstone-paved floors.

"Unicellular exocrine salivary cells with elongated duct-ules is a plesiomorphic condition," says Bill, a long, long way from Tom McGinley's property. But the memories of prairie wet-lands remain, for in a fashion typical for someone who began his work wading in Martin Bay Pond, Bill ends his thesis with an unnecessary appendix on external anatomy and natural history. "Although these observations were not the main focus of this thesis," he writes, "they are of value in explaining some of the similarities and differences found in the feeding apparatuses of these leeches, and help in understanding the function of the salivary cell/proboscis complex."

He'd gone to the big city, done the big city science, but carried along a rationale found in a pond. The pond was a place where kids could find animals. The animals were ones a kid could love and depend on for a lifetime. But Martin Bay Pond was gone. My witching sticks hang in my city office. More than once, I've thought about giving them to Bill. Instead, I think, I'll just show him how to make some and use them himself.

11
Darklings

He was in good company, with a beetle craze sweeping the nation.

— A. DESMOND AND
J. MOORE'S *DARWIN*

Ralene called the first one "Herkimer" and let it ride around on her shoulder. Thereafter they all became Herkimers, these giant, smelly beetles stepping along the paths or across the roads just at sundown, alone, obviously on business of some coleopteran type, pushing their rear ends up in defiance of danger. Did they know what great companions they made? Probably not. But Ralene was never one to pass up a potential pet, especially if that pet also served as an object of her curiosity. Had she actually known a person named Herkimer, someone who preferred the darkness over glaring daylight, who kept to himself even in a crowd, a very large man who couldn't fly like others but who always seemed to be going somewhere important? I don't know; she never told us where the name Herkimer came from. Maybe it was one of those inspirations that simply appear, spontaneously, out of experience, memories, and impressions; an ideal, if not pre-destined, combination of character and name.

The second Herkimer that came down the sandy path got taken apart. Ralene picked up a pair of scissors and started cutting into the abdomen, crunching through the thick, hard, and brittle back which actually consisted of the heavily sclerotized outer wings, fused permanently together. No sooner did she puncture this almost turtlelike shell than a pungent smell wafted through the room, making her eyes sting, permeating her cloth-ing, sticking in her nostrils, there to remain through the evening and into the next morning. I wouldn't try to eat more than one of these, she thought at the time, putting herself in the place of a badger or skunk.

The smell is caused by toxic chemicals with formidable names such as 1,4–benzoquinone and monoalkyl-p-benzo-quinone. Most beetles of the family Tenebrionidae make such compounds, and since many tenebrionids are stored-products pests, bulk-stored cereal grains can become quite contaminated, especially so when stored for long periods under less than ideal conditions. Benzoquinones are known to kill bacteria, fungi, mice, and sperm, as well as inhibit cell division, increase mu-tation rates, and stimulate allergic reactions including ana-phylactic shock, asthma, rashes, keratosis, venous sclerosis, jaundice, anemia, blood in the urine, and maybe even cancer. British dockworkers, who handle large amounts of cereal grain products, are said to be especially prone to dermatitis. Some studies of the distribution of stomach cancer suggest the affliction is most commonly associated with starchy foods. There is no absolute proof, of the type ideally required by a jury, that grain beetles and their secretions cause dermatitis and stomach cancer. There is absolute proof, however, that we live our lives in almost constant contact, much of it direct, with insects. A few of us, however, choose to live this way.

Herkimer II's smell may have protected him from a skunk, but it didn't protect him from a prying mind. As Ralene continued to cut and gag, she found truly giant cells—one-celled animal parasites—looking for all the world like elongate rice grains, living in Herkimer's intestine. For one more time, the results of this completely chance encounter were predictable: The smell be damned; she would study these animals that lived inside of other animals. This moment of decision, reached without much hesitation or forethought, set her on a life's path. The instant Ralene's scissors sliced through Herkimer's intestine and the elegantly attenuated parasites spilled out, a human being was transformed from a student into a professional scientist. From that time on, the trappings of student life—poverty, frustration, part-time waitress job—were only a transparent shell, a thin disguise easily dismissed by anyone who, at some time in the past, had experienced a similar moment of revelation.

"Just don't rub your eyes," Ralene warns.

Herkimer's "real" name is *Eleodes suturalis,* the specific epithet referring to the fusion of the anterior wings—the elytrae—into a hard covering. He and his ilk appear in late afternoons, after the sun sinks low enough to cast shadows in the grass. The world of prairie darkling beetles is one of solitude, blackness, evening. Something happens when the sun drops down; a signal spreads through the tussocks, along the rodent trails, and seeps into the spaces between sand grains. Herkimers everywhere hear this signal, or feel it, and the unseen waves jiggle nerves, sending a message to muscles: Walk the sandy soil, go somewhere. But where? This question is not an easy one to answer. We followed a beetle once, to see where it went. It walked and walked, hesitated, turned, kept walking, left the path and headed off into the grass, climbing and stumbling over fallen stems, came to an open space, crossed a road, and followed a ditch. The light faded. We couldn't see so well, and lost this Herkimer in the dusk.

But in the morning, Herkimers could be found under dried cowpies, or under boards tossed alongside the road, along with several of their relatives. Members of the beetle genus, *Eleodes*, are found throughout the High Plains. Their larvae live in the sandy soil and spend their days eating grass roots. On a bright warm morning, not only Herkimers, but also several other *Eleodes* species can be found hunkered down in a knot beneath a cow pattie. They all have parasites similar to those found in *E. suturalis*, and of course that observation makes you wonder whether they pass the infections around among themselves, sharing gut-dwelling one-celled animals in the same fashion they share the darkness beneath a pile of dried cow dung. And of all the fine properties of cow manure—rich nutrients, moisture, the company of ants and termites and spiders, and darkness—you can't escape the feeling that Herkimer and his kin treasure the darkness over everything else. When you lift up the cowpie, you have only an instant to catch beetles before they scurry into the grass. But *Eleodes suturalis* is by far the largest local member of the darkling family, and the slowest, and so, the most easily caught. Nevertheless, you can't help wondering what you're missing by letting the others get away.

For the next month, Ralene spent her early evenings collecting Herkimers, intending to take them back to the city and raise them. The attempted domestication of wild animals is a highly educational pastime in which one usually fails at starting a colony but succeeds at knocking a giant block off a human ego, namely one's own. But in the beginning, Ralene had every reason to suspect Herkimer beetles would breed in the lab. *Eleodes suturalis* is a member of the family Tenebrionidae, and his cousins thus included some of the most successfully domesticated beetles known to science. Many of the stored-grain pests are tenebrionids. Textbooks are filled with ecological principles derived from experiments with *Tribolium confusum*, the confused flour beetle. Your grandmother's forgotten box of cereal is filled with these very beetles, too, and their ease of culture, coupled with their speed of reproduction, is what makes them attractive subjects for theoretical studies of population dynamics. Another member of the darkling beetle family, *Tenebrio molitor*, the yellow mealworm, has served as lizard and toad food throughout the world's zoos, pet stores, and research laboratories. Scientific libraries are filled with literature on the rearing, embryology, diet, temperature and moisture requirements, life histories, sex determination, parasites and diseases, pathology, species identification, attraction to aromatic chemicals, physiology, and biochemistry of *Tribolium confusum*, its close relative *Tr. castaneum*, and *Te. molitor*. With such a background of cooperation, tenebrionids of all kinds ought to be easy to domesticate.

So Ralene took Herkimer back to the city at the end of the summer, along with another dozen of his kind, thinking that they would mate and have baby Herkimers. She even envisioned a cottage pet industry, picturing kids at school wearing live Herkimers on their shirts, Herkimer shows featuring the biggest, fastest, and most lustrous of these giant beetles parading around in a ring, and even Herkimer veterinarians, larval and pupal nurseries, little "coming out" [literally] parties. But Herkimer proved to be very unlike his relatives when it came to reproduction. He tried, of course, hoisting himself, elephantlike, upon the back of a female, remaining there for hours, pushing his rear end down to hers in what looked for all the world like beetle sex, unashamed, natural, even expected, but not, as it turned out, ultimately successful. Oh Herkimer did *his* job, apparently, and the object of his attentions did hers, too, apparently, for eventually eggs appeared in the bottom of the wheat-bran filled shoebox that was their city home. But none hatched. Herkimer and his mate lived long and active lives, chewing on lettuce, drinking sprinkled drops of water, nibbling at potato slices, producing prodigious numbers of parasite spores. They also spent a goodly part of their time at sex and left behind many eggs. Eventually, a single larva appeared in their shoebox home, but it died soon afterward. Obviously, Herkimer was more tied to the High Plains ecosystem than Ralene realized.

As the months turned into a year, first Herkimer and then the others, passed away, and Ralene chalked up the experience as one more in a long list of typically humbling ones a field biologist comes to expect as his or her due, if not outright reward. After all, an inflated ego tends to be a dangerous trait from which one might want to be saved, and when it comes to deflating dangerous egos, there are few activities that equal the practice of ecology, especially the ecology of infectious organisms. Ralene wanted to study the lives of the animals that first captured her interest—the giant one-celled parasites that live inside Herkimer beetles. But before she could study those lives, she had to get *Eleodes suturalis* to breed in the lab, which it simply would not do. Ralene was never one to be afflicted with the ego disease, however, so rather than continue to force the issue with Herkimer, she relegated him to the realm of nice but uncooperative pets, and turned her attention to grasshoppers and the animals that live inside them.

Three years later, she looked back on Herkimer with a certain amount of fondness. Her grasshoppers wouldn't breed in the lab, either, and one of her species wouldn't even stay alive for more than a day or two. By this time, Ralene had established herself as the consummate wild invertebrate mom, regularly flexing her nurture muscles with a roomful of fifty-gallon marine aquaria, and thus *knew* she was good at the animal care business. Whatever problems these High Plains insects had were decidedly theirs, not hers. She wrote her masters thesis and left town to pursue a doctorate elsewhere. And, she left behind, knowing full well what she'd done, a universe of mystery about the lives, and parasites, of Herkimer beetles and their relatives.

Not long after Ralene left, I got a letter from a young man named Rich. He'd gone to a university in a nearby state to work on his own doctorate, but had decided that what he really wanted to do was study insects, meaning *whole* insects, instead of insect genes, and if not whole insects, then whole something. At a relatively young age, he'd reached a personal conclusion, actually a rather philosophical conclusion, that living organisms are emergent properties and thus must be studied as such. As a result, he'd come to view physics and chemistry as a means of understanding subatomic particles and chemical reactions, not as a means of understanding the lives of whole plants and animals.

We exchanged the usual letters about mutual interests— of which there were many—and employment opportunities for biologists—of which there were few, unless of course, you did what Rich did not want to do, namely study physics and chemistry but claim you were doing biology. He decided to gamble on his future regardless of the consequences. If I have to, I can always sell computers, he said one day. Something in the tone of his voice told me he'd do everything humanly possible to avoid getting a job selling computers. So I offered him a problem, in fact a rather large problem that we'd been trying to solve for years.

"Rich," I said, "if you can solve this problem, then you'll have access to a *real* problem, a true test of your abilities that could make you world famous in a very short time."

The problem was this: domesticate a microscopic animal. That is, make some parasites behave on schedule and on purpose, in the lab, the way they behave in the wild. The behavior Rich had to control was, as is so often the case with wild animals, reproduction, or more properly in this case, early embryonic development. The problematic animals, of course, were those living inside Herkimer and his relatives. In nature, as well as in some colonies of related insects in the laboratory, the infections maintained themselves with apparent ease, which meant that the parasites were undergoing their embryological development quite readily without our help, thank you. On the other hand, in order to do the truly meaningful experiments, Rich had to produce parasite spores on schedule, i.e., on his schedule, not theirs, and under his choice of circumstances, not theirs. If he couldn't manipulate the infections, and guarantee his supply of spores of known origin, then his laboratory colonies might as well be out in the Sandhills along with every other beetle that attracted Rich's attention.

So, like Ralene had done, Rich set about to domesticate not only Herkimer, but his cousins, namely the other species of the genus *Eleodes* that lived in the pastures surrounding places like Dunwoody Pond and the Nevens well tank. In the lab, Rich already had the common mealworm *Tenebrio molitor*, known even by the ancients and described by Linnaeus in the middle 1700s. If he had to, Rich would choose to study the parasites of *Tenebrio molitor*. But like the computer selling job, he would exhaust all other options first. Such is the power that Herkimer and his relatives hold over the initiated. Real biologists, it seems, study wild animals; the *Eleodes* species qualified as wild; something like *Tenebrio molitor* that appeared as if by spontaneous generation in the dust at the bottom of your grain bin—that had been known well even by old Linnaeus—did not.

Here is a photograph that I seem to take over and over again: Alone, holding a net, sometimes an insect net, sometimes another kind of net—an aquatic dip net with a strong handle, heavy steel hoop and sturdy bag—and often carrying a white porcelainized pan or a gallon plastic jug, a young person walks along the edge of a pond, or through the knee-high grass, head down, his or her full attention on the ground. The morning is perfectly still and hot. The horizon is at first impression completely flat, but the more you study this scene, the more that horizon seems to resolve itself into a nested series of gentle low undulations whose colors mix and merge into purple a dozen miles to the east. The young person is the only vertical element displayed against this horizontal backdrop. But the net is held loosely, at the diagonal, a subtle hint of impending action.

This photograph is the iconographic lonely search of a budding scientist in his habitat. The unshaped horizon—read future—always surrounds, ultimately dominating, the scene, and the focus on the ground symbolizes the fact that the planet itself—read universe—is the final source of guiding questions. And the net held loosely? Hold too tightly to your instruments, demand that they always be ready for a single task, and you've committed yourself to a single line of intellectual endeavor. Out walking in the pasture, one person against a flat horizon, you want to hold your net loosely.

And so one of my early photographs of Rich shows him alone, on his knees, in a vast sea of mixed grass prairie, nothing behind him but a flat line and empty sky. His face is low, almost down in the ground, and his hands are hidden. In this photograph, the young scientist is completely isolated, totally immersed in his search, surrounded by the context of his chosen work, as physically strained in his intellectual endeavor as Rodin's *Thinker*, and completely oblivious of the workaday turmoil that you know because you live in it yourself, which lies heavy like a giant trap just beyond the frame. This particular photograph is one of my favorites among a large collection of equally metaphorical pictures of young scientists at the beginning of their careers. The picture is a favorite because of what I know it does not show, namely the action only an initiate can see. An ecologist looks at my work of art and instantly understands the symbolism. But if you've been to the fields for Herkimer beetles, you also see the cactus spines now penetrating Rich's fingers as he grabs—everywhere, in rapidfire succession, for a hundred of Herkimer's cousins hustling off through the blue grama and stem litter.

The cactus spines are a warning, although Rich doesn't know it at this stage of his development, that Ralene's experience with Herkimer is about to be repeated, but on a much larger scale. Rich, too, wants to study the lives of giant cells that live inside tenebrionid beetles, and he carries along some rather formidable instruments, namely a rampant curiosity and almost insatiable need for complexity. It's his characteristic fashion to have decided to study the parasites of not only Herkimer but also those in the several other species of *Eleodes* running frantically, and usually successfully, away from his now cut, scratched, and well-spined hands. Rich is about to find out, if he can, what we were missing all those years before by letting the others get away.

His decision also is spurred by a bit of deceiving luck: Herkimer's parasites will carry out the desired development virtually on demand. The problem I envisioned as being so difficult was not a problem at all, provided you had a supply of wild darkling beetles. The key word in this story is "wild." Rich could have been a doctor, you say, or worked for a drug company making pharmaceuticals to ease our pain, or gone into business selling something, instead of setting himself up for such a grand and meaningless failure. No, he couldn't. This man was born to chase beetles. A dozen miles from the nearest civilization, having studied enough bovine manure to know the optimal sizes, textures, forms and degrees of moisture for finding *Eleodes*, Rich is living the only life his destiny has allowed: collect bugs and study them. And if you end up with a handful of spines, then that's just part of the job.

How do I know these facts? Because my letter from this young scientist was not my first contact with him. Years earlier, he'd been enrolled in one of my very large freshman classes. He was majoring in entomology, out on the agricultural campus, largely because if you majored in one of the ag departments, you didn't have to take a foreign language. That particular decision of his self-guided educational trajectory turned out to be a correct one: A dozen years after the choice to avoid French or German, Rich bought computer programs that would translate the scientific papers he then needed to read so desperately. Somewhere, perhaps everywhere, academic scientists of my grandfather's generation are turning over in their graves, probably cursing in German or French, and depending on the university they attended, Latin and Greek, too. But back to the tale. As the semester progressed, we finally reached the laboratory exercises on insects. The specimens we used were ones I had inherited from at least one generation of predecessors. There were moths in alcohol, their vials stoppered with cork corks and sealed with wax. No expert in insects, even I cringed upon my first encounter with these specimens; a moth in a glass vial of alcohol is an incongruity that stretches the limits of incongruity. A few days after we'd finished that lab exercise, Rich came to me.

"Those are the worst specimens I've ever seen," he said, "they're of no use whatsoever in teaching."

"What do you suggest?" I asked.

"Buy some Riker mounts and cotton, and let me make you a new collection, a good teaching collection," he answered.

Riker mounts are flat cardboard specimen boxes with a glass or plastic panel in the cover, small nails to hold the covers on, and a blank label on the bottom. The specimen is displayed on cotton beneath the window, and although you cannot remove a moth or a beetle, you can when making the preparation, put several individuals in the window, some displayed ventral surface up, others dorsal surface up, and some with wings stretched out or legs arranged especially for study. You can also include enough specimens of a single species on one mount so that morphological and color variations can be studied. But most importantly, unlike insects on a pin, insects in Riker mounts can be handled by hundreds of people, often fairly careless people, without damage.

I thought about Rich's offer for a couple of minutes then took him up on it. The university paid for the materials and a summer's room and board at Camelot. At the end of the summer, he delivered his end of the bargain, namely one of the most inclusive, freshman-resistant, and professionally prepared teaching collections available anywhere in the nation's colleges and universities. It has been years since he took upon himself the task of providing, for future generations, adequate study materials representing his beloved insects. Literally thousands of students have studied those insects. Every semester they get the story of one student who decided, on his own, that instructional materials were inadequate, then set about to correct the problem. I've spent nearly thirty years in the college professor business, and given grades to over twelve thousand students, but so far only one has stepped forward to lay his talents on the line for those who he knew would come after him, also by the thousands. And this is the young man on his knees in the Sandhills, grabbing beetles, a young man not unlike Charles Darwin himself but who, because of his choice of things to study and the date of his birth, has so little hope of finding employment as a teacher. And this in one of the world's most highly educated nations, where colleges and universities abound, and where anyone with the determination can get admitted.

Now, however, from his kneeling position in the grass, Rich hedges his bet against Herkimer by collecting his relatives, too. A spine-filled finger points into the bottom of a gallon plastic jar where a hundred black beetles boil and seethe in a haze of fine dust.

"*Opaca*," he says, meaning *Eleodes opaca*, one of Herkimer's smaller cousins, and one of the speedier. Rich squeezes his hand down into the jar, squinting in the bright sun, his fingers wriggling around in the chaos of frantic beetles. Triumphant, he shakes beetles off the hand and extracts it from the jar, holding a dark angular beast pawing the air with jointed legs.

"*Tricostata.*" *Eleodes tricostata* is a mini-Herkimer, with an almost identical pattern of burnished black over broad deep burgundy stripes; *E. tricostata* is quicker than Herkimer, but just as smelly.

"*Fusiformis.*" *Eleodes fusiformis* is the smallest, fastest, and blackest of the clan; Rich pulls one out, tries to turn it over for me to see, then shrugs as it struggles through his fingers, slips into the grass, and disappears.

"This one's *Asidopsis opaca.*" He holds up a beetle with a flared and sculptured thorax. Same family, different genus. Although only two of these species are named opaque, the word describes the whole bunch. "The really beautiful one with the wide flanges on the thorax, is *Embaphion muricatum.* But you don't see it so often." I get the sense that Heaven to Rich is an August morning, a pickup truck, Nevens pasture, a flat horizon, and a gallon plastic full of darkling beetles.

Before we head back to the car, Rich washes his hands in the stock tank, not so much to remove the cactus spines—which only an idiot thinks come out with washing, or for that matter, will come out at all—but to reduce his possible allergic reaction to these beetles. On the way back to Camelot, he might have to dig something out of his eye, and if there's one thing you want to avoid after messing around with Herkimers, it's rubbing your eyes. Ralene's warning was not lost on any of us. But she only warned of benzoquinones, not of great ideas that could not be explored because of some quirk of nature, ideas that in their own way were even more toxic than Herkimer's chemical defenses.

The great idea concerned the parasites Rich knew grew in profusion inside his collection of beetles. But before you can understand why this idea was such a great one, you must also rid yourself of the feeling that great ideas, especially in science, are ones that earn money or make human life easier. Business-men and university presidents, and especially businessmen who become university presidents, generally cannot rid themselves of this feeling. In fact, about all Rich would have had to do to convince some businessman-turned-college president that the Herkimer parasite idea was a great one, was to tell him that Herkimers ate wheat seedlings and roots, and that the parasites would kill Herkimers. Herkimers do eat vegetation, including wheat seedlings and roots, as do the other *Eleodes* species; these beetles, and especially their larvae, also are called wireworms and early in this century were considered agricultural pests. But there's no wheat within miles of where Rich washes his hands in the Nevens well tank. Herkimers were out in the Sandhills for centuries before anyone came to the High Plains to grow wheat. No, the great idea has nothing to do with agricultural economics and businessmen-turned-university presidents and the feeling that you have to kill something in order to be of value to society. It has to do, instead, with the unifying conceptual framework of all biology. This great idea has to do with evolution.

The great idea is to domesticate these beautiful smelly darkling beetles, bring them into the laboratory where they will be bred in profusion, learn how to raise them without their par-asites, learn how to infect them with their parasites on purpose, learn how to try to infect one species with the parasites from another species, determine the subtle ways in which the various parasites' life cycles differ, determine whether such subtle dif-ferences are the result of evolutionary events that make parasite transmission easier, and finally to match the evolutionary history of these various parasites with that of the beetle species they came from. In essence, the great idea involves a lifetime of work that Rich is seriously considering trying to accomplish in three years.

He would not be considering such foolishness had the par-asites from *Eleodes* not carried out one task so readily and easily, compared with the parasites I'd suggested he study months earlier, namely the making of spores. The parasites from captive *Eleodes* make spores, thousands of spores, with no help from the scientist involved. You get the impression from studying these organisms that they're not much different from anyone else who lives in the Sandhills; in order to survive there, you have to be able and willing to do your job in just about any weather. Unlike the pampered parasites that live in beetles from warm moist grain bins and your grandmother's cupboards, those from wild prairie darklings were not particularly finicky about where they made spores.

Spores are the transmission stages. Beetles eat spores, the parasites emerge, attach to the intestinal lining, live with their anterior ends buried in the intestinal cells, then break off, float free in the intestine, pair, secrete a cyst around themselves, then pass out of the beetle's intestine with the feces. Insect manure dropped in dry pellets, like beetles', is called frass. Across hundreds of miles of High Plains prairies, darkling beetles dump tons of frass laced with parasite cysts. The cysts are the sensitive stage for those species that are vulnerable to environmental conditions. Within these cysts, the paired parasites make gametes, which fuse with one another in an act of fertilization, then become spores by secreting their own cysts. The final act of sex, it seems, is the one that can be disrupted by the wrong set of conditions. But once spores are made, they blow with the wind and wash with the rain across vast reaches of the continent. If you've driven down a dusty plains road, you've breathed in beetle parasite spores. The fact that you're not infected is at the heart of Rich's great evolutionary problem.

The parasites of interest are all members of the genus *Stylocephalus*, so named because of the pointed holdfast structure (*stylo-*) at their anterior, or "head" (*-cephalus*) end. Their spores are formed in chains that look for all the world like microscopic strings of triangular stones. Each pair of *Stylocephalus* from inside the beetle is potentially the source of at least a thousand spores. These parasites are single-celled animals, and in this regard, are similar to the parasites from damselflies that so captivated Tami and Aris. A quarter-million species of beetles, each with its own species of such parasites, and each beetle species with four life-cycle stages—egg, larva, pupa, adult—that could in theory be infected, is a massive problem in evolutionary biology. One might ask, for example, where all these parasites came from, who were their ancestors, what kind of Darwinian adjustments were forced on them by the places they'd ended up colonizing.

And in a more arcane and subtle sense, one might also ask whether larval beetles can be infected with parasites usually found in the adult, or vice versa. The answer to this question magnifies the difficulty of Rich's evolutionary problem. When a beetle larva enters the pupal stage, during which it metamorphoses into the adult, its intestine is flushed with enzymes that destroy everything and the beetle's internal structure is then completely reorganized. Sometimes the pupa is called a resting stage, but what's going on inside—the building of an adult insect—is anything but rest. Pupation might be the most hotly active "rest" Mother Nature ever devised. The question that bores into Rich's mind, occupying his thoughts even as he walks back to the truck, still habitually watching the ground and assessing the beetle-sheltering quality of cowpies, concerns the way blind microscopic animals see the world. He can't quite shake the possibility that these parasites, among the most successful animals on Earth, animals that travel on the wind, finding their times and places purely by chance, view the Herkimer larva and Herkimer adult as two distinct "species," separated by the cellular upheaval going on inside the pupal case.

Such a possibility is the stuff major scientific reputations are made of, namely a conceptual contribution that alters the way we interpret our observations. When Rich's work is understood in this light, suddenly the arcanity disappears. None of the world's warring factions will declare a cease-fire if he pursues, then confirms, this possibility; nor will shaky economies suddenly heal, racial tensions dissolve into mutual trust and respect, street violence dissipate, tropical forests return to their primeval state, and endangered species suddenly flourish. But people will read his report in a scientific journal, then write postcards asking for copies of the paper. The cards will come with a stamp collection affixed: French, Italian, Australian, Russian, Polish, Brazilian, Mexican, Kenyan stamps. Rich will turn these cards over and over, studying the autographs of his scientific heroes now asking for a reprint, studying the people, plants, animals, and events that other nations deem worthy of honor on a postage stamp, comparing that sense of who and what to honor with the analogous values of his own nation; and come to the conclusion that yes, he is in good company with those historical figures who lived in nations swept up in beetle crazes.

But before all this wonderful stuff comes to pass, he must domesticate some species of beetles as well as their corresponding parasites. And at the moment, he's in possession of thoroughly tame grain bin beetles with very wild parasites, and very wild Sandhills darklings with exceedingly cooperative parasites. He's a person holding two extraordinarily large and complicated jigsaw puzzles, each of which is missing key pieces, and if he is ever going to get those cosmopolitan postcards in his mailbox, he must design the pieces necessary to assemble at least one of these puzzles. Knowing that this task will continue into the fall and winter when snow covers both the prairies and the beetles, Rich peers into the future and tries to anticipate the needs of insects he's never tried to rear. So before we leave Nevens, he collects two gigantic garbage bags full of choice dried cow, bull, and horse manure. An alchemist is always alert to potential sources of secret ingredients.

* * *

"With the establishment in 1915, by the Department of Entomology, of project No. 100, dealing with a study of those insects injuring the roots and germinating seeds of staple crops, the writer undertook a study of the available species of the tenebrionid genus *Eleodes*." Thus begins James McColloch's report on false wireworms, with special reference to Little Herkimers, *E. tricostata*. McColloch is one of a very few scientists who have successfully reared species of *Eleodes*. He managed this feat by spending vast amounts of human resources—his time and patience. He also admitted in print that his efforts may well have been spent for love and joy rather than pride and money; in Kansas, *E. tricostata* appeared to be restricted to native prairies.

McColloch kept eggs in small vials, watched over five thousand of them hatch, and determined that it might take a month and a half for this hatching to occur. He also carefully managed the habitat and diet of the resulting larvae, keeping them in one-ounce tin boxes, starting them off with wheat bran as food, then germinating wheat seeds as the larvae grew. He had to change the soil in the tin boxes every ten days. The larvae took most of a year to grow to full size. Pupation lasted nearly a month. Adults were kept in pint jars. At the experiment station where he worked, McColloch also had one critical item of equipment that Rich lacks: a concrete cave. The cave may have been the key to his success with beetles whose family name, Tenebrionidae, refers to their love, indeed requirement, for darkness, murkiness, and even perhaps, obscurity and gloom.

McColloch used the same techniques with another *Eleodes* species, *E. opaca*, and discovered that they, too, spent nearly a year as larvae. He never got around to doing the control experiments, the major interest of businessmen-turned-university presidents. He did, however, visit over two hundred wheat fields, studying the infestations, and taking land use records. After all this work, he concluded that practices every farmer knows and understands—crop rotation, slightly modified planting schedules, and summer fallow—eliminated significant economic losses due to wireworms. That is, McColloch learned the same thing that most parasitologists know, too, namely that despite all the sophisticated and expensive research on vaccines and drugs, the best defense against infectious disease is still a high standard of living: adequate diet, window screens, clean water, and sewage treatment plants.

Was McColloch's beetle research wasted? No. Seventy years after McColloch published the short tables in whose modest entries were hidden his massive labor of love, Rich finds the papers. Instantly he understands completely what is involved in bringing his great idea into fruition, and for that reason, hedges his bet against wild and microscopic organisms. He'll work on intractable parasites in tractable beetles at the same time he works on tractable parasites in intractable beetles. Pursuing two doctoral dissertation projects simultaneously seems a reasonable alternative to biochemistry lab, even if one of the projects involves beetles lackluster and pedestrian enough to live in your kitchen cupboard.

Back in the city, Rich distributes his various beetle species into shoeboxes, tries to adapt McColloch's methods to his own resources, and stores away for future use the thousands of spores his *Eleodes* have produced. He also begins work on *Tenebrio molitor*, the yellow mealworm so well known to those whose jobs it is to store grain and feed lizards. For over two hundred years, scientists, school teachers, and reptile owners have focused their attention on the larva, the adult being treated as a means of producing larvae. Rich, however, begins playing with the adults, assuming, because no one really knows differently, that the adults will have the same parasites as the larvae. But he also simply likes the adults better than he likes the larva. Emotion, coupled perhaps with a bit of the craze that swept Victorian England, makes him do something we'd never done before: isolate adults, hoping to obtain parasite cysts. Then his background knowledge makes its subtle, but logical contribution to his efforts: He moistens the paper towel in a shoebox crawling with fifty dark and driven beetles.

Throughout the night, his charges make their tenebrous rounds of a rectangular box, dropping frass. In the morning, examining this frass beneath the microscope, Rich sees a gleaming white translucent sphere. Using a paintbrush with four hairs, he transfers the cyst to a plastic dish, which he covers and sets aside in a drawer, again in the darkness. Three days later, he again puts the dish beneath a microscope. Instead of the cyst, he now sees coils upon coils of the most delicate filaments, coils stacked concertina-like, coils spreading over the bottom of the dish. The coils are chains of parasite spores. Ralene used to call these spore chains angel hair. We all smiled when she did that; for a parasitologist seeking to control a wild animal, spore chains might well be a gift from God.

For Rich, because of the chosen insects, and thus the particular parasites involved, these spore chains might represent a gift from God, but they most certainly represent a classical case of scientific serendipity. The moist paper towel, misted out of concern for his beetles, provided exactly the right amount of humidity at the exactly crucial time in cyst development. Good scientists never ignore such events; instead, they immediately exploit them. Without hesitation, Rich again isolates *Tenebrio molitor* in a shoebox, again mists the paper towel, and again sets the box aside overnight. Except that this time, he chooses larvae instead of adults.

A week later, he's solved a problem that evidently baffled science for a century and a half: how to make larval parasites make spore chains at your convenience instead of theirs, i.e., the domestication of a parasite. He's used a shoebox, a paper towel, a few drops of water, insight, and perception to solve this problem. The parasites require exactly the right humidity at exactly the right time—a few critical hours in the life of a spore that lives for years—in order to develop. Once developed, spores resist wind, rain, sun, and a pharmacopoeia of chemicals. But in the lab, spore chains produced on schedule suddenly convert ideas and dreams into reality. The experiments can actually be done. Rich looks down through the lenses into the plastic dish. Wrapped up in those fine coils are talks in front of national audiences, published papers, a reputation, his ticket into the future. It's not supposed to be this easy, he thinks. Then he reminds himself that it's not so easy after all. What's easy is using insight and hunches to supplement your science. What's not so easy is convincing yourself that biology rests more on insight and hunches and organisms that will do things for you, than on big bucks and heavy artillery.

Part IV

Memories and Meanings

*E*very year, a day or two before we were to leave, she'd collect the kids and make them stand in a little group so she could take a picture. Eventually she put those snapshots into a frame, behind a mat cut with small panels, and hung this pictorial history of her family summers on the kitchen wall. When she first started taking pictures, the children ranged from four to twelve. Now the children are scattered into different cities. Once in a while they visit Camelot for a day or two, returning to the hills where they spent those idyllic days, wondering who's living in the rooms that made such an indelible mark on their young lives and older memories. I can't always tell them who's living in "their" rooms. Besides, I say, it's the other way around. Those rooms, that smell of the wooden logs and varnish, the birds and grass and good times, those things live in you, not vice versa. Nothing's permanent, not even Cedar Point itself. Even Monkey Rock's washed away and fallen into the lake, I say, but you still call the place Monkey Rock even though it's no longer there. The only thing that's there is the night sky for you to sit under and talk big talk. In the middle of the winter, when it's miserable and snowing outside, their mother sometimes sits at the kitchen desk and looks at those pictures. No, nothing's permanent except the memories and the meanings. You better make them, and use them, while you can.

12
Conversations at the Rock

But even if intelligent signals came to resemble noise when their sophisticated information content was highly compressed, how would we single them out?

— F R A N K D R A K E

There is a place called Monkey Rock, although anyone who's spent the night there talking and watching the stars just calls it the Rock. Even hundreds of miles away, after years have passed, people who've been there say things to one another like, remember that time we were up at the Rock and we were talking about such and such and whatsisname told us about you know and that girl argued with him and then they got into this big fight? Remember that? Hundreds of miles away, and years later, nobody ever asks which rock? The *Monkey* part is understood. Yes, they say, I remember that big fight. We were up there watching the storm and arguing about science and religion and taxonomy and molecular biology and metaphysics and cognitive psychology and the role of women in contemporary society and the stupidity of politicians, but we never did decide whether God created Earth and all its inhabitants, or whether we evolved out of primordial soup.

And then Mark made that picture of primordial soup. Remember? I remember. It was a magnificent piece of graphics— this big Campbell's Soup can, but with the label sort of unfinished, and the words "Primordial Soup" in the diagonal slash, with all the ingredients listed. What ever happened to that picture? Rich has it stuck back in a properly safe place in the lab. What ever happened to Mark? He went off to Chicago. He could have been a commercial artist, easy. He could have been a real artist. Probably. He was a real artist. Or even a musician, or a writer, or anything. So what's he doing in Chicago? Going to school. He might have his Ph.D. by now. In psychology. So what's he going to do? Nobody's heard from him in a while. He worked on beetles when he was out here. Beetles to psychology. Yeah. Silence. In the silence, nobody questions Mark's decisions.

Speaking of primordial soup, that was a great storm. Made you think of those famous Stanley Miller experiments, you know, where this guy put ammonia and methane in a closed container and hit it with electrical sparks and got out amino acids, and thought he'd figured out how life started. Nobody in my class believed Stanley Miller. He's written up in every textbook. Yeah, I know; we just didn't believe it. Did you try it? No. I don't see how you could not believe it unless you'd tried it yourself and failed to make amino acids. Moment of silence. We were trying to describe the lightning instead, she says, I mean, like, using words. An arc of brilliant searing ultrawhite cuts cracking through the pitch black, an angular ribbon fractured into a thousand, no, a million, short connected pieces spreading, spewing off fine sharp scratchboard webbing, a pool of fire at its feet, a quick series of rumbling flashes before the sky explodes, pounding concussions, echoes off the water, ringing in the ears, the far bank dim blue gray with hard line shadows then gone, disappeared into the night. In the distance the clouds flicker, random codes for thunderstorm, dampened grumbles. Silence. You should have been here last night. Why? Because they were better. That one was pretty good. We give it a seven. But last night there were a bunch of eights and nines. What does it take to get a ten? I guess it has to hit the Rock. Uneasy chuckles from around the cliff. The sound of a can being crushed.

I guess I should have come up last night. Where were you? Reading. Grading your papers. Making some slides. I was tired. Read a book and went to bed. Should have come anyway. Who all's up here? It would be better if the power district would turn off the lights over by the inlet tower. Silence. Everybody wants to shoot out the lights over on the dam so they can see the stars. Jeez, that's something like you might hear from the NRA. Well, yeah, but you know, maybe if they thought about turning out the lights so people could see the stars better, then we wouldn't be thinking about shooting them out. On the other hand, y'know, maybe there's some reason why the lights are on over there. So what kind of reason would there be for leaving the lights on so you can't see the stars, huh? Silence. There are no reasons. The noise comes first, like a rush of tiny tornadoes, electricity in the leaves, then the boom comes with the bolt, shattering, shaking, snapping back and forth over the sky, a shower of sparks along the track afterward, a brief white shadow on a black sky. That was a good one. It's a little close. Maybe an eight. You gotta die sometime. Yeah. It might as well be here, watching the sky. Too bad your last thought was a wish that the power district would turn out their lights so you could see the stars. The lightning would be better, too, if those guys would just turn out the lights.

Did we ever decide why people sit and watch the stars? I thought we answered that question already; I know we talked about it a lot one night. You may not have been here. I've been here almost every night. I think you missed the one when we spent all that time talking about why people watch the stars. You were off in town, or back in the city, or back at camp or somewhere. I think we resolved that mystery. Really? Yeah; we came to the conclusion that stars are the ultimate part of nature that we can't control, so that's why people watch them, and make up stories about them, like constellations and myths, and study them to see if they are dangerous in some way, knowing that you can never reach them, and if you made any kind of contact with anyone out there that contact would be with people who could do things you couldn't do. Yeah, but you can reach them if you want to, I mean through physics and astronomy, and all that instrumentation like radio telescopes. You can get right into them, study their reactions, you know, their temperatures, and figure out their life cycles. You can do all that stuff. But you can't *control* the stars. No, but you can control somebody's ability to watch the stars and wonder about what's up there. You can put someone in a dungeon where nobody can see out. Then you lose all track of time. You can't tell whether it's night or day. That happens sometimes back at school. No; really. I go in real early while it's still dark, and if it's in the middle of the winter and I have a late meeting or something, then it's dark when I come out of the building. You work in a dungeon? Laughter. Sometimes it seems that way. Silence. And even when it's dark, all the city lights hide the stars, so you're in this dungeon made of light. A dungeon made of light. I'll have to remember that. Silence. Rumble on the horizon. Orange flashes, sputtering, backlighting clouds, wind picking up. Flickering yellow orange in the north, behind the thunderhead.

Remember whatsisname? He said he watched the stars so he could count satellites going by. Remember how he showed us how to find satellites, and how then we were able to find a bunch? Yeah. But we never could find as many as he did. He was paranoid. He thought they were spying on him. Yeah, but he was wrong. They weren't spying on him; they were only reflecting ideas back down to Earth from some far-off place, and those ideas were getting into his head and making him sick. Yeah, but they were only ideas that we were putting out ourselves, you know, humans were sending those ideas into space, so they weren't really ideas from space but ideas from Earth. Ideas from far-off places made him sick? We thought they made him sick because he was a person who didn't want to have ideas from far-off places in his head. So when those ideas got inside him, they conflicted with whatever was already there. That conflict was what made him sick. That's what we decided, anyway. Quick flash; disappointing snap and rumble.

Why didn't he just keep the far-off ideas out of his head? I mean, he didn't *have* to let them in, did he? He couldn't keep them out. You can't ever keep out ideas. They find a thousand ways to get inside your head. They're infective. Sometimes they're virulent. They make you behave in all kinds of strange ways. Yeah, like looking for worms inside fish. Laughter. Sound of a can being crushed. Sound of a can being opened. Some ideas are a lot more infective than others. Like the idea that we're made in God's image. Now that's an infective idea. It's pretty easy to convince someone he's made in God's image. Then that idea goes all through his body, and takes over his whole mind, and distorts his view of the rest of the universe. It gets into the back of his eyes so that when he looks up at the stars he doesn't really see stars, but instead sees himself. They say birds use the stars to migrate. Is that right? Silence. Nobody knows if birds use the stars. But if they do, then that tells you we're not the only ones on Earth who use the stars to organize our lives, right? So what do you think, that birds look up at the stars and get infected with some wild ideas and end up believing they're made in God's image? Who knows. You'd have to be a bird. But not just any bird. A smart bird. Like a crow or a magpie. Bluejays and chickadees are pretty smart, too. I'd rather be a magpie.

Never can figure out why people don't come in off the lake. Maybe they just want to watch the storm. From the middle of the lake? Silence. Flicker in the north; echo flash in the south; wind; spatters.

Do you believe in reincarnation? What does reincarnation have to do with seeing yourself in the stars? Well, maybe not much, but it has plenty to do with being a magpie. A cigarette glows, brightly, briefly. Yeah; plenty. If you can see yourself in the stars, you can see yourself in anything. You know. Beetles, birds, fish, coyotes. If I were going to be reincarnated, I'd rather come back as a coyote. As compared to what? Well, like a carp. Laughter. Uneasy laughter. Everyone knows what happens to carp. They get stabbed. They get cut up and thrown up on the bank. Then they go flop around until they're dead. People hate carp. They don't even treat carp like fish. Carp *are* fish. Yeah, but because they're *carp*, they don't get treated like fish. So even though you're a fish, you're not a fish because you're a carp. I bet fish don't know the difference. Sure they know the difference between themselves and some other kinds of fish. Yeah, but they don't hate other fish just because they're carp. Well maybe not. Anyway, I'd rather come back as a coyote.

You think coyotes aren't hated? Well, yeah, they're hated, but not so much like carp. I think they're respected. I think they're hated because they're respected. Silence. The only reason they're respected is because you can't catch 'em on dough bait, or snag 'em with a hook. If you could catch coyotes with dough bait, they'd be hanging around rotting on all the fence posts. Instead, you got old boots. Laughter. Why do they put those old boots up on the posts? They put them up there because everyone else puts them up there. Laughter. You never want to throw an old boot away, right? I mean, you've worn 'em. They're a part of you. Even though they may not be good for much, they're still comfortable, right? So you just can't throw 'em away. So an alternative to throwing away an old boot is to stick it up on a fence

post? Why not? Silence. Yeah; why not? No reason, I guess. Then
you can drive by every day and see how long your old boot lasts
out in the weather. They last a long time. Silence. Whose boot
is that up by White Gate? Silence. Some former Director, probably; trying to act like a cowboy, trying to see how long he'd last
out in the weather. I think old boots last longer than old admin-
istrators. Laughter. More spatters. Go up to the car? Not yet.
Silence.

I remember one time when a friend picked up a baby coyote
that had been hit on the highway. It wasn't hurt too badly, so he
took it in to the vet to get some shots. He was going to keep it
as a pet. A pet coyote? There's no such thing as a pet coyote.
Yeah; the vet asked him where he'd found it, then said he should
have run over it again. That's about as bad as being a carp, isn't
it? When you're a baby coyote that's been smacked out on the
highway and even this vet who's supposed to be your friend says
maybe you should have just been smacked again? So what happened to the coyote? It got its shots and then ran away. That
figures. There's no such thing as a pet coyote. That's why if I
had to be reincarnated, I'd come back as a coyote. I don't want
to be anyone's pet. Ooooh-aah! Well, you know what I mean! No,
tell us what you mean! Laughter. I mean I'd like to sort of be in
charge of my own life. Then get out of physical therapy. A
chuckle from off in the dark. Hey, this is the Rock, not the office.
Sorry. But I mean you've got to be able to talk about anything
up at the Rock, right? Right. So what's wrong with physical therapy? Nothing. What'd you make in organic chemistry? A. What'd
you make in genetics? A. What'd you make in calculus? A. Like
high school football players' knees, do you? Ooooh-ahh! Laughter. What the hell's that supposed to mean? Offended. What's a
person who makes A's in organic chemistry, calculus, and genetics doing spending her life with anterior cruciate ligaments?
Hey, this is too serious. I'm coming back as a coyote, and you're
coming back as a beetle. Now *that's* serious! Laughter. Flash in

the clouds overhead. Rumble. Coyotes eat beetles. Coyotes eat anything; that's why they're successful. Broad beam, bent, angular, fixed in place on the south horizon, arc light beam, flickering, over and over again, on, off, signal; stop. That's what it would look like. What? The message. The message? The message from outer space. Silence. Dampened hammering from the south horizon. More wind.

You could come back as a thunderstorm. Thunderstorms aren't alive. Growling, objections, chuckles. Well, I mean, they're not alive in the same sense as a coyote's alive. Yeah, maybe, but a coyote's not alive in the same sense that a thunderhead is, either. So who's to say which one is really alive? They're both alive. You know, if you could come back as one thunderhead, theoretically you could come back as any number of them. Think of that. Even after you're dead you could form up and come back to the Rock and dump lightning and rain on people like us, and entertain us. And make us think about how fragile we are, you know, like when you'd send this gigantic bolt down to smack the morning glory. I think we've found the proof that you can't come back as a thunderhead. Why not? Somebody would have smacked the morning glory on purpose and knocked out the lights. Maybe no Cedar Point alums have decided to be reincarnated as a thunderstorm. Maybe no Cedar Point alums have died yet. Silence. Crack, split, flash, splinter flash, pop; wind; rumble then concussion. Close. Maybe there's someone trying to hit the morning glory after all. Soft laughter. Uneasy laughter. Wind. More spatters. And more. I'd hate to think you couldn't aim a little better than that after you'd been to Heaven and been reincarnated.

Car headlights cast beams out over the lake. Sounds of gravel crunching, engine shut off, dampened music. Music off. I guess he wanted to hear the end of the song. It's a stupid song. May be a stupid song, but that guy's making a million off it. Doesn't change anything. People make millions off of all kinds of stupid stuff. Hoola hoops. Why is it that someone always mentions hoola hoops when the idea of making millions off of something really stupid comes up? Chuckles in the dark. Well, what's more stupid and commercially viable than a hoola hoop? Can you still buy one? No. It's not the hoola hoop so much as it is the *properties* of hoola hoop. Hoolahoopiness. Laughter. Now what could we think of that would have a high concentration of hoolahoopiness? Molecular biology? C'mon; this is the Rock. Laughter. No, seriously. Biotech? Sure, why not? Let's see, what could we make real quickly that would earn us all a million dollars? Pet coyotes. No, seriously. If you could find a way to take the wildness out of coyotes, they'd make beautiful pets. Then they wouldn't be coyotes.

Silence. Wind. Some dust. Big drops. Rumble. Louder rumble. Flicker.

Car door slams.

Yeah, but they might still look like coyotes, and you could call them coyotes, and nobody would know the difference. I mean, they'd buy 'em, right? You could start a fad. Maybe so, but they still wouldn't be real coyotes. I don't think people want real coyotes. I don't think people want real anything. Sure they do. No they don't. Look at television. So what? Everyone knows it's not real. That's just the point. They want to sit on their butts and *think* that what's happening in front of them is real. So they'll curl up by the fire with their TV and genetically engineered fake coyote and think they've got the real thing. They will have the real thing. Not a real coyote. No, but they'll have a real fake coyote. Silence. Softly. There's enough real fake stuff nowadays. Keep the goddamn genetic engineers away from the coyotes. There's enough real fake stuff nowadays. Silence. Rumbles. Flashes. Pop spark streak split shattered arc over the south shore. Strobe light, ghost strobe light, dark spots in your eyes, whitecaps. I don't think you have anything to worry about. Silence. Rumble. Wind. I don't see those lights out on the lake anymore. Wonder if those people got in. More big drops coming faster. Muffled conversation. Sound of feet slipping on the steep path down from the parking lot.

Whereya' been? In town. Yeah? Doin' what? Doing laundry. Everybody has to do laundry. Yeah. Eventually. Which one did you go to? Across the street from the grocery store. I went there a couple of days ago and put my clothes in a washer that didn't work. Really? Which one? The yellow one about third or fourth from the end, over next to the dryers. Never put your stuff in a yellow washer. Why not? Bad luck. Yellow washers never work. The green ones are better. Yeah, but you gotta put your clothes in a yellow dryer because there aren't any green dryers. Yeah, but you ever really look at those dryers? I mean they've got all this burned stuff up by the temperature controls, like there's been some kind of little fires in there. Why would a dryer have a little fire that spews out the temperature control knob, huh? Who knows. Why would they all do that, huh? Nobody knows how washers and dryers work. Now wait a minute, there's somebody knows how washers and dryers work. Well, maybe, but not in this town. Now wait a minute, I bet there's lots of washers and dryers in this town, you know, in private houses, and somebody's got to fix them. Okay, so maybe there's somebody in town but he sure doesn't go down to that laundromat across from the grocery store very often. Or if he does, maybe he just hates to work on yellow washers, but loves the green washers. Maybe it's not a "he"; maybe it's a "she." Hey, that's possible in this day and age. Yeah. But in this town? Maybe.

Silence.

I never had any problems with those washers and dryers. Maybe they like you. They're showing mercy. Machines show mercy? Washers and dryers aren't complicated enough to show mercy. You have to be complicated to show mercy? Well, if you're a machine. I mean like some computers show mercy, and some cars show mercy, and those are pretty complicated machines. But washers and dryers aren't complicated. But you show mercy on machines, you know, like you'd never abuse your microscope or anything. A microscope isn't really a machine. Of course a microscope is a machine. If a microscope isn't a machine, then what the hell is a machine, huh? You don't even abuse your forceps. Not abusing my forceps is the same as showing mercy on them? Well, they're sort of the same. So what is a microscope if it isn't a machine. Well, maybe it's made like a machine, but a machine is something you use, and I think a microscope eventually uses you. Silence. You think your car doesn't use you? Got to show more mercy on washers and dyers of the world or they'll quit just like your old car. I think they have those little fires coming out of the temperature controls because there've been people in there who don't have any mercy. Meet all kinds at the laundry. That's for sure. Rumble. Spatters. That's for damn sure. More spatters. Flickers.

Silence.

Y'all going to sit out here in the rain?

Maybe.

So what's goin' on in town? Besides laundry? Yeah. Not much. Went into the Sip for a while. Met some guy who knows you. All the locals know him! Laughter. No, this guy wasn't a local. Really? What'd he look like? Sort of beat up. Said he'd been here ten years ago. That could be anyone. He asked me if we still went up to the Rock. So what'd you tell him? I told him it depended. On? Well, you know, I didn't want this guy following me. So what'd you tell him it depended on? Whether it was going to be a good sunset or a bad one, or a good storm or a bad one. You know, that kind of stuff. There aren't any bad storms. That's what this guy said. What else did he say? He said there weren't any bad sunsets, either. Well you wasted your time worrying about whether he'd follow you. I didn't invite him. You don't have to. Anyone says there's no bad storms can find his way out here. There are bad storms. If they tear up your wheat or drop a tree limb through your roof then they're a bad storm. I wasn't judging storms by the damage. Besides, he already missed the sunset. Silence. Wind. More spatters, faster spatters, bigger drops. Flash. Rumble.

Maybe he saw it from somewhere else. Then it wouldn't be a sunset at the Rock. But if this guy's been to the Rock, then he knows that whatever sunset he's watching is also a sunset at the Rock. Silence. Heavy. More spatters. How wet you gonna get? I don't know. Wet, though. Well, not real wet. But wet. It's only water. It's not like you haven't been wading around in the damn river all day. Different kind of wet. Wet's wet. Silence. No, I don't agree with that. I think that when you want to be wet, and don't mind being wet, and have some real business being wet, then you don't notice it. But if you don't have any real business being wet, then you want to be dry. So being out here on the Rock is dry business. Silence. For some of us. We'll go sit in the car. It's not quite that wet. Faster spatters. Always pushing it, aren't you. Rain never hurt anyone.

Brilliant instantaneous white fire searing light and trail of sparks ripping tearing splitting explosion; echoes off the canyons; echoes in the clouds; light echoes in the north; rain. Web across the sky, crackling web, sparks, gone; boom; trailers; echoes. Feet slip on mud. Car doors slam. Smell of soaked jeans and sweatshirts. Mud splatters on windshield. More mud spatters, then steady blowing pounding water. Might have got real wet out there. Yeah. Lightning fractured through glass. Thunder pumps inside the closed car. You think I'm not real wet? Quit breathing, you're fogging up the windows. Gotta breathe. No need to breathe if you can't see the storm. Wind. Hard wind. Flash. Car rocks. Gravel blowing at the doors. Pop. Oh hell. Pop. Pop. Pop pop; ping. Hail. I knew this guy from California who came out here one time and always parked under a tree. Really? Yeah. The only tree was half a mile down the road, you know, down by the ranch house, so he parked down there. Why? 'Cause he was afraid of tornadoes. Silence. Sad silence. Nobody told him that cottonwoods are the first to go. Gravel against the metal door. Pop pop ping; pop; ping ping. I hate it when the hail breaks out your rearview mirror. Yeah. Me too. Rearview mirrors are hell to replace. Roar. Blowing gravel and water; sheets of water down the windshield. Wind rocks the car. Wonder if those people made it in off the lake. Silence.

Ought to be some way to see this except from inside. Only way you can see a storm is from across the lake. That's not the same. I mean there ought to be some way you could see a storm from inside it. No, right, but still, you can't see what you really want to see when you're away from it. That's what I said. Silence. But then you can't feel it. Why don't you get out if you want to feel it. No, you don't understand; I want to see it like you see it from across the lake *and* like you see it from inside all at once, like seeing and feeling it at the same time. So why don't you just get out? I don't want to feel it that bad. It's like those physicists who can't see two aspects of their subatomic particles. Anybody ever get blown off the Rock? What do you mean, and live to tell about it? Well, yeah. I mean there are people who've gone over Niagara Falls in a barrel and lived. That's just a story. It's a metaphor. No, it's a parable. No, it's a metaphor, like when you say it's like going over Niagara Falls in a barrel. That's a simile. Metaphors and similes are similar. But I think going off the Rock in a storm is not quite so bad as going over Niagara Falls in a barrel. Would you accept going off the Rock in a storm in a barrel? As what? As equivalent to going over Niagara Falls? You mean as a metaphor, or as a simile? Or as equivalent to not being able to see your storm any better than the physicists can see their electrons? Silence. I don't know.

Hiss. Rattle. Roar of rain and gravel. Sheets of water down the windshield; sheets of water down the door window. Wipe it off so we can see. Leaves a mark on the inside. Yeah, but wipe it off anyway so we can see. Nothing out there but water and electricity. Yeah, but that's what we came up here to see. I thought you came up for the sunset. Silence. Sometimes sunsets last 'til morning. You just call it a sunset, but if it's a storm, then it's still a sunset. You call it a sunset when you go up to watch it, but if it turns out to be a storm, then you remember it as a storm but you still call it a sunset. All you're seein' is water running down a windshield. Knocking on window. Somebody's out there. Let 'im in. Knock knock knock knock. Hey, Doc! Tell 'im to go around to the other side. GO AROUND TO THE OTHER SIDE! Sheets of water and gravel. Wind. Arc light piercing strobe bam! ka-bamm! GO AROUND TO THE OTHER SIDE! Let the poor guy in! It's blowin' from that way; tell him to go around. GO AROUND! Knock knock knock; other side. Water. Sheets of water in the car. Gravel in the car. Wind through open door. Enormous wet blob in the car. Slam. Hi, Doc; saw one of your kids in town doin' laundry and thought you'd be out here. Who is this guy? Introductions. Might as well open the windows. Please don't.

Where you been? What do you mean "where"? Just "where"? I've been working. Do we get stories? Stories? Yeah, old-timer stories. Tell 'em some old-timer stories, Doc. Chuckle. Rain. Silence. Tell 'em that story about the guy who spent all that time watching the stars looking for satellites. He thought the Russians were coming. Same satellites now; no more Russians. I told 'em that already. Isn't that something that you can get your mind all twisted from lookin' up at a Russian satellite then live long enough so that there's no more Russians? I mean, what do you do with all that twisted mind you got from worryin' about the Russian satellites, huh? There's still Russians. No, I mean *real Russians*. Like in the Russians Are Coming! Russians? Yeah, those kind of Russians. If you're gonna' get your mind all twisted from watching satellites then you got to go find yourself some more Russians. But they might not be Russians. Of course they're not Russians. So what are they? Somebody good enough to put up satellites. There's nobody playin' Russian now that's good enough to put up satellites. Physical therapists? Silence. Slow rain. Rumble. Physical therapists put up satellites? No, but they're coming. "The physical therapists are coming!" Yeah. I mean like they're coming to get your smartest kids. Silence. Spatters. Gonna quit. Muddy down on the Rock. There's a patch of light way over in the west. I can't see it. You can see it from out on the highway. Silence.

I don't think the physical therapists are coming to get the brightest kids. So who's doin' it? I don't know. I think it's someone else. I think the physical therapists are just doing their jobs and not worrying about the brightest kids. Silence. You're right. They're worrying about high school football player knees. Silence. Rain's stopped. Basketball players, too. Basketball players? Yeah. The physical therapists worry about basketball players' knees, too. I don't want to think about basketball players. Really? Why not? It's summer. Roll down the window. Get this fog out of here. Rain's stopped. Run the wipers. Bzzzzzzz; thunk; thunk; thunk. You need a new blade. There is a clear spot in the west. We'll get to see some stars after all. Right. Maybe there'll be some physical therapy satellites. One of these days you're going to be very old and maybe have a stroke or some kind of an injury and need a physical therapist. I'm not going to worry about that. You're not worrying about gettin' old? No. There'll be plenty of physical therapists and they'll all be the best and the brightest we had to offer. We didn't send 'em off to war. Silence. We sent a few off to war. But we won that war real quick. Rain's stopped. Rumble way off in the east. Roll down the windows.

Silence. Match flares; pungent smoke through the car. Amazing people can keep their matches dry walking through the rain. I didn't walk all the way out here. Hitch? Some guy goin' to Lewellen. I got this plastic bag. I keep my stuff in a plastic bag. Silence. What would you do if they didn't make plastic bags? I guess I'd have to get a goat's bladder or something like that. Soft chuckle. Silence. Cigarette flares; smoke drifts through car, out open window. Sqwaak. Baby owls, over in that hole in the bank. They nest there every year. They were nesting there when I was out here. That's been a long time. Twelve years. Twelve years is a long time. Twelve years goes like a minute. Doesn't seem like twelve years. They may be nesting in that hole in the cliff for twelve hundred years. Maybe. Orange spot's wider. Going to clear off. Dark shadow, silent black shadow, across the windshield; gone. Owl's out. Sqwaak. Gotta feed them babies.

So what's new, Doc?

Nothing much. Same thing. Study some animals. Go on down to the river, get some fish, cut fish. Get out of the way of your graduate students. Yeah. Stay out of the way of your students and they'll make you famous. Well, maybe not famous, but happy. You were always a happy guy, Doc. Not always. Well, most always. Not even most always. Enough. You were always out on the river, Doc. I remember you, and maybe it was that Steve kid, always out on the river after those little fish. Red dot brightens; sound of breathing; eyes in the faint glow; matted hair and beard; piercing smell of used tobacco drifting through the wet darkness. We were out on the river a lot. Easy to be happy out on the river. Then you guys would be up all night long cutting, with your eyes stuck to those microscopes. Those must have been good days, Doc. Ever hear from Steve? How's the river, Doc? It was dark when I came across. Have I missed anything?

You mean in the past twelve years?

Yeah. Have I missed anything in the past twelve years? I mean, have you guys found out anything new that you didn't know twelve years ago? What's happened, Doc? What have you gotten out of twelve years on the river, huh? I mean, you're still here, but where's Steve, huh? What'd he *get* from being out on the river all that time? Huh?

I don't know what Steve got from the river.

Tell me a story, Doc. Tell me a story about the river. Have

you guys gone to the river this summer? Tell me how the river's doin', Doc; how's the old Platte gettin' along?

Yipping, yip yip yipping, howl song.

Coyotes back in those canyons.

Tell me a story, Doc.

Once upon a time there was this road, called the Road to Roscoe.

I know that road.

Once upon a time there was this road. . . .

13
The Road to Roscoe

Try to feel *the curves & sur-*
faces, and sweep them in freely
instead of engraving them. Get
my point?

— LOUIS AGASSIZ FUERTES

1

My mother died at the age of forty-eight, having battled
cancer for nearly a third of those years. I hardly remember her
when she was not ill, sometimes gravely so. When our first child
was born, we took this baby daughter at the age of nine days to
see her grandmother, who mustered all the strength she had, sat
up in bed, and held Cindy while my father took a picture. The
photograph was a 2 × 2 slide. When it came back from being
developed, the left side of it, the half with my mother's face, was

rather dark, but the other part with the baby, was light. The flash attachment on my father's camera hadn't been exactly synchronized with the shutter, so although the picture was in focus, and clearly showed a woman on her death bed holding a baby, one was in shadow and the other in light. We had no opportunity to take another picture. On the day the slides came back in the mail, my father was busy with funeral arrangements.

That series of events happened thirty years ago but they still stand out sharply in my mind because of the central role time played in the little drama. It seems appropriate to me to begin a discussion of prairie rivers with a story of my mother's encounter with her first grandchild. To stand in the Platte River, to feel the sand move from beneath your feet, is to understand at last, deeply and powerfully, that the passage of time like the passage of sand and water, is completely, totally, and finally, irreversible. Furthermore, the moving grains of silicon tell you that natural sequences of events, once set into motion, continue of their own accord. The continent of North America will not stop washing away to the sea simply because you stand upon it. No amount of love, desire, care, hope, no infant granddaughter, could have stopped my mother from dying. Once my father pressed the shutter, the machine was set into motion according to Newton's laws and Einstein's assertions about levers, forces, springs, and light, producing a photograph of enormous symbolic power: a dying woman in the dark, holding a baby in the light. And once that image was placed into the memory of a person who would one day write essays, nothing could stop it from being recalled and used to make a point.

My mother always kept a book of Robert Burns' poetry beside her bed, nestled in among the forest of pill bottles. It was a special edition of some kind, bound in leather that had an indescribable living feel. That book felt like one of those newly hatched, silver dollar-sized, soft-shelled turtles that you seine up about once every three years from an oxbow along the Platte

River. The paper was high quality flimsy; when you turned the pages, they made the same kind of sound as dried cottonwood leaves blowing over a gravel island. Periodically my mother would demand that I read something from that book, usually for my own edification, and more than once the lines were from Burns' famous poem about a louse. I suspect the opening line was identical to the major question in her mind about her son—*Ha! Whare ye gaun, ye crowlin' ferlie?* Eventually, and inevitably, we'd come to the verse about seeing ourselves as others see us.

I got the message, of course; who wouldn't. But the book contained another message, a disguised one, that depended on my having seen the poetry in another book and in another circumstance. Burns' original words were not the same as in the Americanized version I'd read in public school. Somehow to "see oursels as ithers see us" had a meaning that "see ourselves as others see us" lacked. The original spelling as a physical printed series of letters and punctuation marks in a particular typeface meant "others" in a way that transcended the dictionary definition. The fact that my mother put those words in front of me over and over again meant that a bit of timeless insight was being made a part of a pattern, almost as if she sensed that when she was gone, I would miss being reminded of how I "looked" to the rest of the world, and so would seek those words or their equivalent.

On the other hand, perhaps she was also asking that I not see her as most people would—a dying woman in pain long past the requisite reconciliation with her condition—but as a person who, in her own way, would lie there alone and mock the spectre of death. Her book of Burns deflated whatever heroics the medical profession with its surgery, intravenous needles, chemotherapy, radiation and hormones, may have appeared to perform. I never saw her take a pill; but I saw her read the poems. Do not let people view you as meat, she was saying to me, symbolically, to be cut upon, stabbed, fed toxins, rolled strapped and helpless on your back beneath a cobalt beam, in order to be kept "alive." No, to stay alive, return again and again to the insight of the ages wherever you find it. Return whenever you can as long as you have the strength to do it, to a source of images that nourish your mind. So she gathered her strength and opened the book with the softshell turtle cover and cottonwood-leaf pages.

Small wonder that upon discovery of a prairie river's powers of personal renewal, I thought immediately of her nightstand with its amber bottles, their implications of technological sophistication, and the book that was her real medicine. She would go to Burns and the poetry would come to her. And I would go to the South Platte near Roscoe, and the river would come to me from the mountains. I would stand barefoot to let the sand wash from beneath my feet and think of Robert Burns. I could not bring myself to see the river as "ithers" saw it. The ithers wanted to dam it somewhere near the Rockies, so Denver could have green lawns, nice golf courses, sewage flowing in its city system, laundromats, carwashes. I had this vision of an engineer coming into our home and taking that book of poetry from my mother's bedside as she lay sleeping, drugged by some latest pill that was supposed, in one shot, to cure what ailed her.

2

The modern history of the Platte, like that of many rivers in the west, is a grand experiment to find out just how much water a river "contains." That is, we would like to know the upper limits of its ability to gather water. "Contain" can refer to the instantaneous total volume from the headwaters to the mouth, or in another sense, to the amount that is ultimately determined over long time periods by the physical geography of its watershed. In other words, we want to know the volume of water that defines existence for a river, in this case, the Platte, with its branches, the North and South Platte Rivers, originating in the Wyoming and Colorado mountains, respectively.

"Existence" implies all of the human interactions with this flowing stream, including the history, the migrations, diversions, songs, stories, art, music, inspiration, indeed anything in which humans are or have been involved from the first time one laid eyes on it, after having walked from the Bering Land Bridge to the Central Plains, until the last time one studied its braided sandy tracks from an airliner miles above. The volume of water that defines such existence is equal to that traditionally considered "wasted" by developers and engineers, unless it is used by some human for a specific purpose directly associated with the generation of income. Flushing your toilet counts; writing poetry usually does not. Hopefully, and in some respects strangely, however, birdwatching has started to count, and that fact alters the equations by which engineers calculate the waste.

The experimental aspects of river use, as all experiments in social and political behavior as well as in medicine, including cancer treatment, are derived from the mental processes of prediction combined with manipulation. Humans propose to carry out a physical act and predict the ultimate consequences. True scientists first write or think a null hypothesis—nothing more than a highly utilitarian research tool—which says, in essence, that nothing will happen as a result of their acts. Crucial to the overall activity, of course, is the maintenance of a "control," i.e., a set of unmanipulated materials with which to compare the manipulated ones when the experiment is finished. After the experiment, if there is a difference between controls and manipulated materials, then the null hypothesis is falsified and the scientists have made progress toward understanding a natural phenomenon. By showing that one prediction is false, they've inferred the truth of another. With these kinds of short, sometimes sideways, steps, science stumbles its way toward a rational explanation of nature.

Societies have carried out massive experiments of the kinds just described, although the crucial element of deductive science—the control—is typically missing from such manipulations. But the null hypotheses for these studies can be written easily: A meltdown at a nuclear power station will not increase the incidence of cancer in a local population; the clearing of all tropical forests will not alter global climate; no DDT-resistant mosquitoes will evolve in areas sprayed with large amounts of DDT for several years; the invention of the computer will have no effect on the private lives of citizens; the damming of the Platte will not alter the physical structure of that river three hundred miles downstream where five hundred thousand sandhill cranes must spend the month of March feeding in order to nest successfully in Canada three months later. For some reason, these grand experiments do not seem to have the same aura of cleanliness that surrounds the more esoteric ones done for the ten millionth time in freshman biology lab.

What is also missing from such monumental tests of rather naive null hypotheses is the opportunity to repeat them. No power can truly turn the clock back to that day in 1962 when my mother first held a grandchild. The photograph cannot be retaken; we cannot recover extinct beetles gone up in Brazilian smoke, eliminate computers from our daily lives, or dismantle the knowledge of how to build nuclear weapons. The sand cannot be returned to the mountains, nor the water to the Rocky Mountain storm clouds. And when the wasted river is finally saved— sucked dry and sold—then the engineers' predictions will have been adequately tested. We will find out what happens to a society that sells its metaphor.

3

The town of Paxton is marked, as most are in this part of the world, by a service station at the interstate exchange, a grain elevator you can see above the treeline, and a bridge over the South Platte River. In town are Swede's Cafe, a door that used to open into Cheetah's Lounge, and Ole's Big Game Bar. I turn off at the Paxton exit, thinking about one year, back in New Jersey, when I got on I-80 to go west, and another year when I got on I-80 in San Francisco to go east, and how both trips eventually ended at the same place—in Paxton. I am driving and dreaming, refusing to look at the river as I cross the bridge because of what I know I'll see, and anticipating what I'll find ten miles down a two-lane highway at another town, this one called Roscoe. I am testing my own null hypothesis: There is no difference between dreams and reality.

In town, car parts bang beneath my feet as the ancient station wagon bounces over the double rail crossing. A hundred yards to the north is an awning made of rough-hewn lumber, shading the front door of a place where a local man displays the results of a lifetime's labor—the killing of wild animals. In my dream, Ole's Big Game Bar is out of business, abandoned. I park, walk in the blistering bright heat of late July up to Ole's window, and shade my eyes to peer within. I used to bring students here; now most of those kids are doctors, attorneys, or teachers. When we first started coming to Paxton, a giraffe's head nailed to a knotty pine wall symbolized adventure. Young people wandered from room to room, studying photographs: Ole with professional football coaches, Ole with actors, Ole as a member of the Paxton High School basketball team, Ole with arms full of ducks and geese, Ole with his foot on a bear's neck. Kids also studied the heads: zebra, elephant, the antelope series, as well as a row of dusty stiffly strutting pheasants and grouse on the shelf above the bottles. You could see the questions on the kids' faces then:

What was it like in Africa? Were you scared facing that bear? Then twenty years passed. The next generation had different kinds of questions: Why, someone always wanted to know, would a trophy hunter kill a dik-dik or pay to have a hornbill, labeled "vulture," mounted in a bar?

I was never able to answer except by saying that at one time it must have been all right to kill dik-diks. Indeed, my African traveler friends tell me dik-diks are about like cottontails, in Africa. I'd pass along the information. Then we'd talk about the fleas and tapeworms that had been wasted. None of the students I took to Ole's would have killed a wolverine without then looking for its fleas and tapeworms. These conversations never seemed to end satisfactorily, but in trying to answer their own questions, the students would eventually arrive at some understanding of why the large brown glass eyes in the little head were staring out over the dance floor. At one time it was not only all right to go to Africa and kill a dik-dik, but more significantly, to mount it on your tavern wall beside a photograph signed by a professional football coach. The antelope became subordinate to the history lesson. Ole's time, like wasted Platte water, had flowed past Paxton, never to be seen, or felt, again.

Today, in my dream, Ole's is closed, abandoned, empty except for the heads stacked around the floor, furniture piled in corners, pool table covers ripped, a beer bottle on the bar. A few light rays spear the dusty darkness; a deer mouse, tail held high, skips along a baseboard. The place still smells the same, but without the jukebox and tourists, the building itself makes sounds—some pops, a moan of boards, starlings scrambling in the ductwork. Outside I hear the wind blowing, hard, howling around stripped car bodies and among the pipes that issue from the elevator towers. The large windows rattle in their frames with the wind, and I can hear the fine staccato peppering of river sand blown against the glass. A half mile south of town, between the tracks and the interstate, the South Platte bed lies dry. But the restless sand will not wait; if it cannot ride the ripples, it will ride the wind. On the sidewalk beneath the awning, a man with a cane tells me Ole's ain't open no more. I say it doesn't matter; neither is the river; I was just curious. He looks at me like I'm a fool. As I drive away, I see him staring at my license plate and in that instant everything becomes clear to him. I'm from the city. There's no way I would have known Ole's was closed, or the South Platte dry, unless I dreamed it on the way west.

My second null hypothesis is that with the passing of Ole's, the community in which it resides will not be altered in any perceptible way. Ole's was born of a time when the African continent was a symbol of pure adventurous exploration. Ernest Hemingway seemed to speak for the American male psyche, Rider Haggard's fantasy underwrote our wishes for a stylized Dark Continent, and Ethiopia was for Louis Agassiz Fuertes, the greatest of wildlife artists, a verdent mystic paradise that came close to validating Haggard's fiction. It was into this limitless and alien wilderness that Ole set out to kill meat and collect trophies which he brought back to Paxton and nailed to his tavern wall. Then a man who'd never been beyond his barbed wire fence except that time they went to the state high school basketball tournament could sit in Ole's on a Friday night, nurse a beer, look around at what Ole had brought back to Paxton, and feel a little more worldly because Ole'd shown them that being from Paxton couldn't stop you from killing a dik-dik in Africa. So by analogy, being from Paxton couldn't stop you, or your children, from doing anything you, or they, wanted to do.

But in my dream, identifiable blocks of Paxton's lifetime end, as did the ones on my mother's bed, synchronously. A human life of forty-eight years, a gestation period of nine months, an hour's drive to the city, and a sixtieth second inside a camera, all had come to their conclusions at almost the same instant. But now, in my dream, the geological epochs since the separation of Africa from what is now South America, generations since Rider Haggard wrote *King Solomon's Mines* and Hemingway wrote *The Snows of Kilimanjaro*, the years since Ole pulled the trigger on a dik-dik, the months since the last violence in Soweto, the days since a state legislator representing an agrarian people locked into an arid land introduced a bill, in brazen violation of fundamental principles of survival, to sell eternal water rights for quick cash, the hours it had taken me to drive to the physical setting in which my ideas had always flowed like the ripples and darted like the shiners, and the fraction of a second it took me to recognize that Ole's Big Game Bar, that monumental symbol of exploitative glory was gone, all ended synchronously. In my dream. After all this time since the Mesozoic, Paxton was about to find out that the unlimited wilderness was just one man's passing fancy.

With this conclusion, still in my dream, I get in my car, leave Paxton, and turn west toward Roscoe. And as I drive in the searing afternoon heat, the corrosive south wind whipping cornstalks on my right, my thoughts turn to another building made by another man to hold his treasures.

4

My father died at fifty-nine at Thanksgiving, his tumor having been diagnosed in March. He was a collector of succulent plants, as proud, I felt at the time, of the cactus he'd started from seeds as he was of anything else he'd accomplished in life. Cactus seeds don't just sprout like beans. In nature, the plants often spread vegetatively, one of the most prolific being the ornery prickly pear, whose pads break off, sticking to deer, antelope, dogs, cows, horses, pickup truck tires, jeans, then travel as far and as fast as the carrier can take them. But my father had a knack for starting cactus from seeds. I know nothing about this business except what I learned from cleaning out his garage after he was gone. That experience taught me that a cactus collection, unlike heads in a big game bar, is sustained only with an enormous amount of luck, care, skill, soil mixtures, gravel types, special water, and an artist's sensitivity. For it was not enough that my father's collection stay alive; no, it also had to bloom and its seeds had to sprout and grow into seedlings. And for this requirement he decided, on the day he was released from the hospital in the spring, that with his last remaining strength he would build a greenhouse.

With the meticulous detail typical of his approach to everything from his consulting geologist business to the mixing of a martini, he planned, and singlehandedly constructed, a small greenhouse with temperature and humidity controls. The one question that has remained in my mind is whether he actually planned the project so that it would consume precisely and exactly every minute of his last few months. Within hours of the afternoon I pushed his wheelchair out to the greenhouse so he could simply sit and enjoy what he'd done, he went to sleep never, really, to awaken.

Fifteen years later my sister and I tossed the contents of his soil buckets over the back fence into an Oklahoma creek bed. The labels on these recycled antifreeze cans were lettered in strokes so close to Helvetica Light Italic typeface that I found myself staring, as I always did at his printing, wondering how a human could smoke and drink so much and still exert such control over his own hands. The soils we finally threw away were artificial versions of the Arizona landscape, mixed to both physical and chemical perfection according to some secret recipe. By this time most of his cacti were gone, the important ones dispersed to other collectors, the hardy ones distributed among his surviving relatives.

The greenhouse is obviously a metaphorical lens that brings into view and focus, our traditionally hamfisted approach to the domestication of surface waters, which is why these thoughts come to me as I drive the few miles to Roscoe. It was very clear to my father, and to those around him, that if cacti were to be made, by human hands, to bloom and germinate, then those hands had to be the sensitive ones of the artist, directed by a brain that did not overlook for convenience sake, either the details of an ecosystem or the tales told by land and cacti themselves. So I toy with the idea of building a greenhouse to hold a river. This is not a strange thought; a woman who chooses Burns for her bedfellow and a man who makes cactus bloom would understand.

5

A gentleman who owns a parcel of land adjacent to the river near Roscoe has given me permission to use the place as a site for teaching and research. He also has given me a key to the gate, the first and only such privilege I have ever been offered. For that reason, the key is symbolic of an understanding between two people who really don't know one another except through their values. I didn't ask for permission to drive my all-terrain vehicle over his property. No, one of my students asked for permission to study toads, and I to seine for small fish.

Then one day I decided to take an entire group to Roscoe. They were young, interested in one another instead of minnows, and didn't behave properly. They splashed in the river without first looking where they were walking, talked too loudly about the wrong ideas, laughing and joking about what they'd done the night before and what they were going to do later that night. I told the student who studied toads that I was angry; all that splashing and loud talk showed lack of respect for minnows and for the worms that lived inside them. He agreed; he'd been thinking the same thing.

The next week I decided we would repeat that day at the river, this time with the proper attitude. We would study parasites properly, I yelled, and pounded my fist on the table. You wouldn't play grab-ass in church, would you? What in the hell makes you think it's okay to do that in the river? Those were the only words I knew to truly express what I was feeling. I'm sure none of them had ever heard a person get angry because someone else talked too loudly while a killdeer screamed and ripples whispered. So we went back to Roscoe. Their eyes were sharp. They heard the sounds of water, read the turtle tracks, felt the individual sand grains washing to the sea.

Weeks afterward, one of them said, you never give up, do you? But all the time I was thinking it was time to give up, that my feelings about the river were mine alone, and that I should savor them while I could, the same way I did our children when they were growing up, instead of trying to force those feelings onto others. Let the politicians and pork barrel engineers and lawyers fight over the Platte, I thought, tired. They're going to kill it eventually anyway; why stand in their way? But then one of the students who'd been taken to Roscoe and told to shut up and listen to the river's stories said, you never give up, do you? And I looked at her and said no, you don't ever give up.

6

At some time in the past there had been a tremendous flood. The waters had ripped up cottonwoods, dead and alive, and sent them down the river toward the ocean. This historical event is written in the distribution of driftwood. How much force would it take to move a stump that size, you ask, standing in a muddy pool a half mile from the water. A river that could carry a cottonwood tree five feet in diameter could float a battleship. But such a river, in this land, can also go dry overnight, leaving the trunk abandoned in a ditch. Then floodplain brush grows up high enough to hide the dead tree, fungus eats away at cavities where wrens go looking for nesting places, the thick bark separates from the underlying wood, the stump turns white, and a human comes along, finds the tree lying on its side, and is fascinated by the patterns revealed by the roots of a giant cottonwood thrust upward by a flood.

Midstream driftwood is always masculine: young soldiers ripped from their neighborhoods by a flood of events over which they have no control, lying sprawled and twisted on a beach; midlife executives suddenly stranded on a gravel bar, their limbs still green but withering, reaching for the sumacs; a long straight trunk, waterlogged now, half submerged, around which the shiners play—he never read a book or listened to a symphony and is proud of it, but he did his job the way he was supposed to do it, then he retired. Now he watches the minnows play; he's a fan; sometimes his team wins. Compared to the stump, the midstream log is simple; the wrens do not come to him; he has no complexity in which to hide.

I cannot take my eyes off the upended mass of roots stranded back in the brush. They are all the tales ever told by an old woman looking back on her life. Behind her is a mile of floodplain; a thousand, maybe ten thousand, young cottonwoods are clattering in the hot wind. If she could see them she might say, I was like that once. Then the shadows on the writhing tangle tell what she was able to do with her opportunities. A root twists to the left—she was allowed to grow in that direction. Another bends sharply to the right—something got in her way and she had to change course. One curls back on itself—she felt she was going in circles, not accomplishing anything, but even though she's past that stage in her life the coil shows she's "been there," so she knows what it's like to not have a direction. I see hundreds of these tracks through time; they are each unique; there will never again be another cottonwood with exactly these roots. Their twists and turns may have seemed to be problems, years ago, but they held her upright for decades. Now, at rest, her sustaining roots, hidden for so long beneath the sandy soil, are exposed to the July wind. A man studies them, and sees in them a symbol for a rich and varied life.

7

A highway sign asking us not to litter the beaches around a local lake, reads: Leave only your footprints. I think of this sign as I study killdeer tracks, trying to reconstruct the behavior of the bird that made them. But this exercise is a trivial one, made so by the fact that I've watched too many live plovers. So I phrase the problem another way: What does a person have to do to eliminate prior experience, or preconceived notions, from an interpretation of some new experience? On the one hand, our history is our identity; a personal past shapes thinking and vision in a singular way. "Unique" is an adjective with unsettling implications; you happen only once in the history of the universe. On the other hand, experience is also a prison. I know how killdeers behave, so am forever denied the challenge of discovering this behavior from their tracks. How much more fortunate are the paleontologists, I think, who never saw a dinosaur but get to study their trails in rock that was once Mesozoic mud.

My mind plays with the highway sign, and with the sanity of the person who made up the slogan. I'd like to take some footprints, not leave them, from the river at Roscoe. How wonderful it would be to send killdeer trails, instead of letters, to those I wished to influence. Then the recipients of these bits of living South Platte would wonder why they'd received bird tracks. Those who'd watched a killdeer would know the answer instantly: Someone was telling them how hard it is to see the world from a new perspective. A gentleman by the name of Marc Reisner once spoke to this problem in a book as elegant as its title: *Cadillac Desert*, an exposé of America's brutal relationship with its most beautiful assets, its rivers. From the rather routine and outright lies about cost/benefit ratios, to the stupid engineering decisions made in the face of geological evidence that everyone associated with a project knew and understood, *Cadillac Desert* is an epic tale of how point of view dictates morality. Water that flows toward the ocean, proclaims this point of view, is simply wasted. And waste is immoral.

Beneath my feet is the physical evidence of a plover running through the mud, stopping in places, backtracking. Here's where it stopped, bobbing its head, screaming its name; here's where it chased a beetle; here's where it stood and preened. How do I know these things? I don't, but the scenes I've watched unfold in the past come back as the "truth" that I "know" just from noting depressions made by three small toes. These scenes are pressed into my brain, permanently, just like the word "wasted" is embedded in our collective mind. *A prairie river shall not be wasted.* It can be killed, but not wasted. It can be taken from society as a whole to supply the wants of a few, but it cannot be wasted. Whatever we do, we must not waste the Platte. But we seem not to have learned the lesson of the killdeer tracks: You don't have them unless you let the bird run free.

One day near the end of summer I returned to Roscoe to get a gift for a friend. I carried a large can and a sack of plaster of Paris. Finding a series of tracks that told me a story I wanted to read over and over again, I mixed the plaster using river water, and poured it over the prints. The first cast turned out so beautifully, I made three others. Killdeer tracks to take home. They now sit like a time bomb behind my typewriter. Some day, in some arena, in front of some audience, this bomb will explode. That day will come when the Platte is finally dry, when Denver's sewage is washed away with water that once flowed past Roscoe, when the cranes no longer darken Nebraska's March skies, and when the July winds blow sand against the boards on Ole's windows. Eventually someone will want me to give a speech—Rotary Club—it happens all the time. Afterward they'll ask where the river went. I'll pull out the plaster cast and say, this is what we've wasted because we let the wrong people tell us what we were wasting.

8

There is a final scene I must describe. It was a year of high water, a rare year in a land prone to drought. When the water is high, the river's physical complexity is submerged, floodplains are under water, and for weeks on end a brown and urgent, heavy, almost primeval, South Platte races toward the ocean. You don't have to wade in very far to discover that a high river is a dangerous river. The speed of the water is frightening, but its force is even more so. Standing at the roaring water's edge fifty feet off the highway in June, your seine slung over your shoulder, suddenly aware of your vulnerability, makes you wonder what the plains must have been like before settlement. It was in such a year that we set out for Roscoe, went into the river as far as we could and tried to catch the fish we were studying, but found the terns instead.

It was in the flood year that the least terns nested at Roscoe and I took my wife Karen to see them. When the water finally subsided enough, and we could walk into the floodplain a few hundred yards, I discovered what the South Platte had done, namely bestowed certain properties on a pile of sand. In most years the sand pile was just another hump on an otherwise flat land. But when the river roared, the sand pile became isolated in the middle of the torrent, half a mile from shore, surrounded by raging water through which no intelligent coyote would swim. There the least terns, called by instincts many millions of years old, came to nest. Drawn by an appreciation of what a rare event this nesting was, I took Karen to see the babies.

I'd noticed the adults first. The air had been filled with terns, hurtling air knives, their calls long fingernails on slate, a qualitative addition to the river, something new, different. Never before in my memory had there been the least terns at Roscoe. A scientist is always alert to strange observations, always suspicious about their implications. On the sand pile was a newly hatched chick, hunkered down in a depression half as deep as a footprint. Having seen the first one, you could then see them all over the dune, single frizzies no bigger than your little finger, crouched in the unmerciful prairie sun. I put my ear down to the chick, listened to its peeps, heard the parent screeching above. Then I went to get Karen. When we returned, I spent a roll of film on the diving, slicing adults, put a tape recorder into the air to catch their sounds, and spent another roll of film on the babies. At some time, I thought, humanity will need proof that the South Platte River near Roscoe is a sacred place. That's when I will play the tape and show the pictures and someone will say, surely Roscoe is blessed for the least terns nested there once.

We stayed away from the river for a month. I did not want to know what happened to terns whose island had suddenly become accessible to humans on foot. When we finally went back, all terrain vehicles' knobby tire tracks criss-crossed the dune over and over again, making circles in places where a person had put a foot down and spun in the sand. The terns were gone, the air silent, the river muddy, sullen, and subdued. I said I was glad this dune was not wasted, that some person got some use out of it. Not long after that day, the gate to the river at Roscoe was locked. A student of mine who wanted to study toads visited the owner, who listened to his request that must have sounded as foreign as a tern call, then loaned us the key.

As with the picture of the sick woman with her first grandchild, or the completion of a greenhouse, this incident with the least terns has that structure in which identifiable blocks of time end synchronously, and the result is an experience of such power that it stays in your mind as a major event. A river is, of course, one of our most obvious metaphors for a human life. The headwaters are often inauspicious, small but beautiful. With the passage of time-water, the river grows, becomes more complex, gains assets, joins with tributaries. The stream is mature when it has the capacity to rage and direct the fates of those around it, and dies when it reaches the ocean, carrying the mud and debris of a symbolic lifetime.

But the river can also be enslaved, shackled, so that it will not be "wasted." But sometimes it can temporarily writhe free of chains. Then the least terns come to it and put their babies in its care, laying them on the glaring surface of its islands protected only by the rushing water. But this primal trust is broken when ATVs grind chicks to sandy paste. The river says, I was not free to give you what you needed after all. Then the terns go into the sky—cloud birds—in search of another sacred island. And when I return to the August Platte, with its hundreds of ankle-deep warm crystal channels and silent clear pools, I cannot escape the memories brought forth by running water and moving sand.

14
The Blue Mustang

Man in his arrogance thinks himself a
great work, worthy the interposition
of a diety.

— C H A R L E S   D A R W I N

Naked Ape

— D E S M O N D   M O R R I S

Imagine a scene: right before dawn, three days after
the summer solstice, not a breath of air, the lake perfectly
smooth, not even a duck foot ripple, and a dense fog lying over
the water. From somewhere comes a soft croak and a splash; you
can't determine which direction the sounds come from, you have
no idea how far away they originate. The light changes so quickly
that even if you have your camera in hand, the decision about
exactly when to take a picture is extraordinarily difficult. You
know that one of the most wondrous and unique scenes you'll

ever witness is passing quickly into just another routinely clear hot morning. You didn't produce this event and can't control it; you feel deeply honored, even privileged, to have been shown the fleeting combination of light and subtle color. Your neural networks fire; their collective action invokes your mind; your mind wonders about the meaning, the explanation, for this fortunate, transcendent experience. The concept, the *idea*, of privilege, of being the humble and in your own mind somewhat undeserving recipient of a taste of grandeur, sweeps through your entire body and stirs your emotions. It's the hand of God at work, you think, the supernatural, omnipotent, Hand of God.

No scientist comes walking by spouting out an impromptu lecture on dewpoint, refraction, angle of incidence, chaos theory, strange attractors, and the probability of observing a foggy sunrise over a High Plains wetland in June. No grubby field biologist who's spent a lifetime outside in the predawn hours comes grumbling that the fog obscures his view of the pelicans. No parasitologist drives by in her pickup truck full of traps with mice destined to be "turned into" tapeworms and fleas; no ichthyologist suddenly fires up an outboard motor somewhere in the middle of your transcendent experience, then materializes out of the fog beaching a muddy aluminum boat, and throwing three hundred feet of sucker-laden gill net at your feet on the sand. No warning whistle from the hydroelectric plant half a mile away shatters the stillness with its declaration that the turbines are about to start.

None of these mundane and ever so human events intrude upon your communication with the forces you cannot control. Instead, you pass the serenely peaceful moments before breakfast in contemplation of your own place in nature, your insignificance compared to the infinite universe, your brief moment on Earth compared to the eternity you're convinced lies waiting in the hopefully distant future. There must be a "reason for it all," you conclude, a reason for your being born in the first place, allowed to develop into maturity in the second, and finally, given the privilege of standing on that silent foggy shore. You share this mental state with all who have ever truly experienced a range of natural phenomena—from thunderstorms, tornadoes, prairie fires, and blizzards, to breathtaking sunsets and sunrises, once-in-a-lifetime cloud patterns, the view from high mountains, and the fresh vibrance of a tide pool in the spring.

By "all," I mean *all* humans who have ever lived and spent time looking closely at the sky above them and the ground at their feet. Your foggy June morning on the shore is something you share with Cro-Magnons, Neanderthals, the irascible British painter Joseph Mallord William Turner, the practical Albert Bierstadt, the properly respectful and thoroughly observant Frederic Church; the Australian aborigine, the modern Eskimo pausing for a moment before blasting the crystal stillness with the firing of his snowmobile engine, and the Chinese peasant flooding his field with night soil. As surely as you share the genes that spell their species, you share their atmosphere, their encounters with the planetary forces they, too, cannot control, their wonder at the power of those forces. And, you share their needs for explanatory myths to reconcile their seeming insignificance with the monumental size of a mountain range or the unerring predictability of a star's rise and set. You in the singular, as well as you in the plural, share all these things with all these people past, present, and future.

You also share a desire for moral guidance when faced with the forces you can't control, whether the product of those forces be a foggy sunrise or a holy war. And most of us turn for advice to the forces that produce the fog, rather than to those that generate the shooting. Religion is the traditional source of personal decision-making help in most societies, and if the information often reported in our daily newspapers is a true reflection of the global psyche, religion nowadays is a growth industry. News stories ranging from those about heated political battles to those reporting armed conflict routinely mention the religious factions involved. More often than not, participants on both sides claim the moral high ground. But insofar as we can tell, this situation has been largely true for much of recorded history.

Whatever else it may purport to do, and whatever else it also actually does, religion tells us how to behave and encourages us to pass judgment on the behavior of others. And depending on the denomination, sect, or cult involved, consequences for disregarding this guidance can range from mild to dire, from abstract to quite tangible at the individual level. The relationship between unsanctioned behavior and punishment is well established; there is little evidence from history, literature, or science, that the human species is comfortable with its individuals' decision-making powers. There is plenty of evidence that we seek to place major responsibilities for our lives in the hands of gods who live beyond the reach of our judgment and influence. This seeking may be as much a human trait as our opposable thumb.

But original biology, done dirty in the field using exotic plants or animals, has a profound impact on one's willingness to use his or her God-given gifts. Objectivity, rationality, perception, and intuition are at a premium in this endeavor. Received wisdom, regardless of its source, quickly becomes suspect. The smaller the animals involved, and the less economically important, the more suspect the wisdom. Even the most sophisticated and complex scientific literature on microscopic animals reveals a long, continuous, and ongoing struggle with the question of how to study such material. In contrast to most religions which purport to teach knowledge and certainty, field biology, if not most science, teaches ignorance and uncertainty. A walk through a marsh with a magnifying glass is a lesson in what you have yet to learn. It's also a walk through a world that is largely inaccessible to humanity, not because most humans don't live near a marsh, but because so few humans have, or even want, the information needed in order to see what they don't know. Ask your friends about the last time they studied the ectocommensal protozoans on *Hyalella azteca*, tried to identify a leech, or used an insect net to pick one damselfly out of twenty thousand. If they're able to describe the experience, then they're members of a highly privileged class: people who've seen their planet through a microscope instead of through the eyes of other humans, some of whom carry their own personal agendas.

It is virtually impossible to study small animals without also expanding your sense of wonder, thus making you somewhat vulnerable to non-empirical thoughts. A quart of Dunwoody Pond contains several kinds of rotifers, forty or fifty species of protozoans, half a dozen species of worms, crustaceans belonging to three or four orders, insects belonging to four or five orders and eight or ten families, as well as the diatoms, algae, and uncountable numbers of bacteria some of these animals eat. A handful of South Platte River mud is not quite so bountiful, unless you collect it from the algal mats in late August. The mats themselves provide a lattice upon which a community equivalent to Dunwoody's is displayed. After a few years of looking at mud and aquatic vegetation under a microscope, you view the next petri dish with the same expectation you take to the symphony: You're going to like it and appreciate it regardless of whether you can understand or play it or even know what the instruments are. Clearly there are places in the world you cannot go. Just as clearly, the world, like the symphony, goes on without you. And if you let yourself ramble mentally, dwelling on your inability to handle a French horn or know exactly where the swallows will be nesting in the middle of the next millennium, you can work yourself into a feeling of insignificance in a hurry.

That the world has been going on without you for several billion years, and will go on without you for several billion more after you're gone, is a simple fact that is not always easy to accept. Yet the evidence for such an age, and for the profusion of life forms that have walked the ancient lands and swum the primeval seas, is unequivocal. The same people who give you nuclear weapons—the physical scientists—give you radioisotope dating methods that tell us the age of our planet. The same people who bring you petroleum—the geologists—bring you a map of Pangea, a theory of plate tectonics, and a tale of drifting continents that explains much of the present geographical distribution of plants and animals. The same people who labor to find causes and cures for cystic fibrosis—the molecular biologists—bring you plausible explanations for the origin of cells with mitochondria and chloroplasts. The same people who bring you commercial hairspray, soap, insecticides, patent medicines, plastic and nylon—the chemists—bring you replicating RNA molecules that cast new light on the origin of life itself.

The nonscientist's naiveté, if not outright purposeful obtuseness, is to accept the bombs, gasoline, and hairspray as science's contribution to their lives, while rejecting the cosmology that is an integral part of the scientific enterprise. That is, it's simply dumb to say: Cure my disease but don't try to tell me where I came from. You don't get the cure without the intellectual medicine. Young scientists who go searching for hard questions to answer don't stop working with their minds when they reach some invisible boundary set up by people who've never walked the muddy cattails. Nevertheless, young scientists as well as old ones, eventually come to realize that they are only single individuals in a massive population, and that their work is but a small contribution to the overall product of humanity. Teaching scientists are more fortunate than pure researchers when confronted with a final analysis of their efforts; a living legacy of people whose thoughts are shaped in part by their own ideas is a truly satisfying accomplishment. But no matter who they are, all explorers eventually ask who they are, what it means to have lived the life they chose, and whether they did the right thing. Their experiences, however, make them aware that the questions apply not only to themselves, but also to their species. If there is any philosophical contribution biology has to make to this discussion, it's a vision of species-level morality that necessarily accompanies our acquired intellectual powers as revealed by our works.

Fifteen years ago, I chose to symbolize the human need for meaning and purpose as well as our desire for moral guidance, with a red-headed woman driving a blue 1973 Mustang. There is nothing special about the car; it was one I'd recently looked at on a sales lot but didn't buy. But I called the woman God, obviously a rather blasphemous act, but one intended to democratize, as well as modernize, the moral guidance she was about to give: *Take only what you need.* At the time, she was talking about small fish, but the message is easily extended to include all of Earth's resources. The two scientists who were given the guidance could just as easily have been representatives of our species. I used the narrative structure of a small parable to convey an idea about species-level morality, which is one of the more fundamental issues of our time, and indeed one from which many of our other problems evolve. The question is: How should *humans* behave as opposed to Americans or Russians or Chinese, and as opposed to Kimberly, Heidi, Robyn, Kim-Thanh, Albert, Kipp, Hoanh, and Trevor (a few first names chosen at random from my most current class roster)? God's advice—take only what you need—was a literary device for connecting individual behavior with the species' moral obligations to the planet as I saw them at the time. "You" can be both singular and plural; "need" has many meanings; "take" likewise has a range of translations from the highly specific and tangible to the symbolic; and "only" is a fairly ambiguous term. The phrase can be taken as advice, as I intended it to be, or a command, which I'm sure some readers would prefer it to be.

If the phrase is advice, the implication is that we have legitimate decision-making power; if the phrase is taken as a command, our decision-making power degenerates into the capacity for violation of a Supreme scenario. No teacher assumes that human decision-making power is a degenerate function of the species. Instead, a teacher assumes, must assume, that decision-making power is a legitimate tool for carving out one's individual time and place in planetary history. We must also assume that such power is a legitimate device for defining the species' time and place in planetary history. In this regard we are little different from Charles Brown's swallows; the costs and benefits of group acts can best be determined, and described, through an assessment of the acts' impact on the individual. Nevertheless, it's the species that survives or becomes extinct on the strength of its individuals' behaviors.

The people in this book have taken what they need from the river and left behind a river relatively untouched by their presence. They've also taken what they need from the prairies, potholes, roadside ditches, cattail marshes, spring-fed streams, and dry grasslands, and those habitats remain virtually unaltered by the thesis problems extracted from them. In addition to their needs, these young scientists have satisfied their wants by taking inspirations and appreciations from the foggy predawn mornings, the sunsets, the midnight constellations, the meteors, and the soaring pelicans. Their ability to take something of substance from a falling star rests upon the fact that they are human beings, i.e., upon those genetically endowed and highly evolved traits they share with Cro-Magnons, Huaorani Indians, and Wall Street bankers. And having studied nature from a perspective more akin to that of the Cro-Magnon than the banker, they likely classify falling stars and foggy mornings among their needs. But my students' wants are not necessarily shared by everyone else, past and present.

Indeed, if there is a trait the people in these pages possess, a trait that marks them as individuals rather than members of their group, it's their wants. And if there is a personal possession that defines their wants, a possession they acquire from the system of American education, a possession that is theirs and theirs alone and often drives their behavior, it's their thesis or dissertation problem. This problem is their blue Mustang. In the parable of the Mustang, the car carries the woman two places: down a two-lane highway south toward a small town called Grant, and apparently, if her announced intentions are to be realized, to South America. She's following a sandpiper. The scientists she encountered offered to take her to Lewellen, instead, Lewellen being another small town, in fact much smaller than Grant, about thirty miles west of where they were collecting fish. Her decision to go to Grant rather than Lewellen symbolized the individuality of exploration. Her road, Highway 61 south, was any road, anywhere, even a mental one made from decisions, and she chooses to go south rather than west. In translation, the parable is almost too transparent. The thesis problem is the vehicle, steered by decisions, driven first to the next small town, then beyond, and if intentions are realized, even to places the highway is not supposed to go.

The sandpiper in this parable, a lesser yellowlegs, *Tringa flavipes*, represents all biological problems, in that it offers observations to be made, but at the same time performs some functions that humans may not necessarily be able to decipher. The bird is also representative of that vast wilderness of everyday biology one finds anywhere—Kansas to Amazonia—that has no immediate political or economic importance. The migration to South America symbolizes the journey one takes in pursuit of any research goal. The difficulty of flying to Argentina on your own wings is easily understood to be the ups and downs and diversions one encounters in a lifetime of serious scholarship. The distance from Grant, Nebraska, to the *pampas*, however, is the distance we must see into the future in order to make correct decisions today, decisions about what it means to be human.

No one actually sees very far into the future, of course. Those who possess certain beliefs about eternity look into the future and see their beliefs, which are not necessarily consistent with the planetary events that dictate the way their children and grandchildren must live. Thus we come to the modern theological paradox soon discovered by any scientist raised in an environment perfused with Western Christianity: God's warning to Adam in the Garden of Eden contains an unresolvable conflict between God-given abilities and God-given commands. "From any tree of the Garden you may eat freely; but from the tree of knowledge of good and evil you shall not eat, for in the day that you eat from it you shall surely die," says God in the second chapter of Genesis. Taken literally in context, "good" and "evil" become ignorance vs. knowledge of one's genitals, respectively. Taken as metaphor, the warning becomes a profundity revealed in retrospect: Science and technology will open your eyes, let you see things you've not recognized and understood before, and bring you into contact with parts of the universe hitherto inaccessible. Taken in a modern context as well as metaphorically, "good" and "evil" become the benefits and costs of a scientific enterprise, respectively. Cast from the Garden, humanity had to go to work. We've been working ever since, using our God-given gifts. The result is art, literature, music, architecture, science, technology, agriculture, medicine, instantaneous global communications, and a photograph of Earth from near space—the benefits—and the ozone hole, burning tropical forests, toxic waste, nuclear waste, smart rockets, and instant global communication—the costs.

The parable of the blue Mustang is intended to define "good" and "evil" for the human species now equipped with modern science and technology, rather than for the individual. "Take only what you need" means do not destroy the planet that supports you. Control your population before it controls you. Don't be stupid. Don't listen to false prophets who tell you not to use contraceptives. Don't be naive about what the planet can sustain in the way of human flesh and suffer in the way of fire, excavation, and garbage. Use your powerful brains instead of your equally powerful herd instincts. Make art, music, science, literature, intellectual freedom and mutual respect, not war, including war upon your Earth, and especially not war inspired by a god. You don't need 10 billion more human bodies on the planet some of you think was made in seven days. You need to learn to read good books and tolerate a range of viewpoints and desires. Only humans make art, science, and literature; ants make war and babies. Only humans search for life in distant galaxies; robins search for worms to feed their children. That's what the driver of the blue Mustang, the individual in control of her technology and using it to go exploring in places we didn't think it would go, is telling us.

I had no choice, of course, but to write her as a beautiful red-headed woman. No one would have listened to just one more college professor spouting arcane profundities derived from a study of obscure animals. Why beautiful? Because as a society, if not as a species, we pass early judgment on female physical beauty, make decisions about it before it grows mature enough to make its own decisions, lay expectations on it, put it in cages to admire, assign it roles that make it a liability, hold it back from places ugliness is allowed to go. That is, we routinely treat beautiful women the same way we treat the planet—use, control, fear, put her on our calendars, admire her from afar, but keep her out of math class, send her to be the nurse instead of the doctor. I felt it was necessary to give a voice to a segment of our society that usually has words put in its mouth. Her beauty is the untapped human potential we ignore because of preconceived notions about its value. Why red-headed? Red hair is a hackneyed symbol for feistiness and independence; her hair color represents rebellion, in this case against the prevailing business world view that Earth is ours to exploit.

And why a woman? If I need to answer that question, you need to read some additional books. She's the half of the human species that historically, along with her children, has paid the price for the other half's grand adventures. Every dead soldier has a mother; every dead soldier's mother is convinced that it is right for her to bear the death of her child in obeyance to a commander-in-chief and that she must keep her mouth shut or be considered a traitor. Every child who cannot get enough to eat has a mother; every child who carries a handgun to school or buys cocaine on the corner has a mother. God is a woman because of that long parade of brilliant young women I could not convince to seek positions of intellectual leadership in their society. The redhead in the blue Mustang is not going to be a physical therapist or an occupational therapist or a nurse. She's not going to spend her life messing with the torn anterior cruciate ligaments of high school football players. She's on her way to Argentina instead, following a bird. We get the sense that when she returns, she may have something to tell us.

And where did her authority come from? Three acts. First, she drove up and got out of her car uninvited. She took the initiative to comment on the behavior of others, not condemning them for what they were doing at the time, but giving them guidance. She knew the scientists were humans and could do easily what they damn well pleased, but her advice—"Take only what you need"—touched a deep sense of responsibility about planetary resources, that they already possessed but whose importance they may not have realized. In a social species such as ours, initiative gives a person authority, at least temporarily, and when that initiative reaches a latent thought and brings it out into the open, the authority is strengthened. We tend to obey commands consistent with what we've learned already on our own. Secondly, she spoke to the other animals in our immediate vicinity—sparrows and the sandpiper, again reminding them of their freedom as well as their "responsibilities." It was time for the sandpiper to go to South America but the bird was lingering, maybe too long in the fall, on the Platte. This authority rested on her knowledge of natural cycles. Again, she was reinforcing what all the participants in the parable already knew: The planet and your genes place constraints on your behavior; violate the constraints and you spend a winter on the Platte, metaphorically as well as literally. And thirdly, she was doing a more audacious piece of biology than the scientists were. She acquired authority from her choice of "thesis problems." She picked one we thought was impossible to do, and she picked it, evidently without a great deal of forethought, in order to satisfy her curiosity. Among people who study nature, a desire to satisfy one's curiosity regardless of the consequences gives a person a great deal of authority among the like-minded. To us, she was embarking on the most noble of endeavors: Follow a piece of the natural world in order to find out where it goes. Period. For no other reason.

Now I'd like to tell you a story that I've not told very many other people. There used to be a place we called Ackley Valley South, a marsh right along Highway 61. When the poison ivy got so bad at one of my other favorite places, Mudhole #1, I stopped bringing young people there on the first day of summer class and started taking them instead to Ackley Valley South. It was not a big marsh, but it did extend to both sides of the highway, was always filled with snails and insects and parasites, and had a convenient and safe parking place nearby. A perfect classroom. Then Ackley Valley South started to dry up. One year we went back there near the end of the summer just to collect before it dried up completely. The water was very low west of the highway, and there were wide mudflats around the edge. The morning was calm; the sun was beginning to warm the air and evaporate the dew. We'd just gotten out of the vans and I was talking to the group, when I heard the call of a sandpiper. From out of nowhere, a lesser yellowlegs appeared. I asked my students to stand still and watch. The bird flew around and around the shrinking marsh, calling. It must have flown in a circle thirty or forty times, passing almost over our heads, then out over the flats, calling. Then it left.

The next year, Ackley Valley South was dry, gone the way of Martin Bay Pond. The year after that, it was hay meadow. Now you'd never know there used to be a marsh right beside the highway, a place where any kid could go to get muddy and educated. There was a time when I would see a sandpiper and feel excited, pleased. Then there came a time when I would see a sandpiper and feel privileged. Nowadays, as often as not, when I see a sandpiper I feel sad. Where have they gone, I wonder, those Dunwoody Ponds, Martin Bay Ponds, Nevens well tanks, Roscoe river beds, Monkey Rocks, and Ackley Valley Souths? And if they're truly gone, then where is a sandpiper to land when it comes into my door looking for places to find *Hyalella azteca*, beetles, leeches, damselflies, minnows, frogs, swallow colonies, worms, and a place to study falling stars? These are the questions I remember in late August, when public school starts, and when my memories of those young scientists who've passed through my life make me look out over an ocean of eighteen-year-old faces and ask: Who's next?

Notes and Acknowledgments

I would like to extend my sincere thanks and appreciation to the individuals who contributed to this book, both directly and indirectly. My current group of graduate students has been especially helpful, patient, and cooperative, although by the time you read this they likely will be gone from the fold. They are, in order of appearance: Aris Efting, Tami Percival, Laura Krebs, Scott ("Marv") Snyder, and Rich Clopton. Aris received her Bachelor of Science from the University of Nebraska–Lincoln in the summer of 1992, then started immediately thereafter as a graduate student in my laboratory. Tami really was in my freshman zoology class and wrote five papers on a fossil sponge. She stayed to do graduate work, received her Master of Science degree in 1992, and went to Texas A & M to study the protozoans that live inside termites for her Ph.D. Laura came to Nebraska from Wake Forest University. (The person who gave her the book was Ray Kuhn, an immunologist at Wake Forest.) Laura eventually ended up doing her Master of Science thesis research on the parasites of two species of minnows. Scott Snyder is a doctoral student at the University of Nebraska–Lincoln. He is originally from Auburn, Nebraska, a small town in the southeastern part of the state where his father was a senior administrator at one of the nearby small state colleges. Scott was an honors student as an undergraduate, went to Wake Forest to get his Master of Science degree, and returned to Nebraska for doctoral studies. I have no idea where he got the nickname Marv. Rich is Richard Clopton, a highly successful doctoral student at the University of Nebraska–Lincoln in the early 1990s, winner of student paper competitions at regional and national scientific meetings, and a computer whiz. Rich received his Ph.D. in the summer of 1993 and went to Texas A & M on a postdoctoral fellowship to study the parasites of insects. Other graduate students mentioned were gone before this book was written.

Rich, Scott (Marv), Aris, and Laura read the complete manuscript, as did undergraduates Jill Anderson, Mike Barger, and Dan Holiday; also J. T. Self, my former adviser, and Jane Geske, co-owner of Niobrara Books and former director of the state library commission, and offered their comments. Tami had gone to Texas by the time the manuscript was finished and saw only her chapter. Mary Lou Pritchard, Ralene Mitschler, Skip Sterner, and Bruce Lang also read individual chapters and suggested changes.

Lee is Lee Hardin, who received his Master of Science at the University of Nebraska–Lincoln, went to Tulane to medical school, and is now a physician in Hawaii. Of the Leech Queens, Mary Ann McDowell stayed to do her Master of Science on the parasites of fathead minnows; she's now an immunology doctoral student at the University of Wisconsin. Midge is Marjorie Gardner, who also went to the University of Wisconsin to pursue graduate work. Liz is the former Elizabeth Larson, now Mrs. Snyder, who worked as an undergraduate in my laboratory. Bill is Bill Moser, who was an undergraduate at the University of Nebraska–Lincoln but went to the University of Toronto for graduate work. Ralene is Ralene Mitschler, who was also mentioned as a magician with marine aquaria in my book *Vermilion Sea*. At the time of this writing, she was a doctoral student at Kansas State University, about to receive her degree, and applying for various jobs. Two that are not specifically mentioned are Tim Ruhnke and Mike Ferdig, although their research and efforts in the lab contributed significantly to many of the ideas that permeate these pages. Tim now has his Ph.D. and is working on shark tapeworms at the University of Connecticut; Mike is at the University of Wisconsin studying the resistance of mosquitoes to filarial worms.

The minority student in chapter 5 is Chris Rhodes from Omaha. In response to an editorial query, I traced him down, using an old student phone directory. He remembered the described events as clearly as I did. After reading the chapter manuscript, Chris decided he'd rather be known than anonymous, thus the acknowledgment by name. I greatly appreciate his endorsement of my use of the incident as book material.

The local ranchers and residents of Keith County also have been exceedingly generous with their time and property. Duane Dunwoody, who lives northwest of Keystone, Nebraska, has been a good and lively friend to me and my students for a number of years, always welcoming us to his pond; he also has provided a place for Titus to live since the accident. The Sillason family has allowed me and my students to use their well tank, overflow pond, and adjoining pastures for teaching and research for a number of years. We are truly grateful for this courtesy. Buckhorn Springs Ranch is now owned by the Haythorn Land and Cattle Company; the Haythorn family has always let us use their land for teaching and I am deeply thankful for their generosity. Tom McGinley is part of the McGinley family that has owned ranch property in western Nebraska for several generations, although much of the original holding has been sold. Tom operates a convenience store, campground, and laundry on the Keystone Road near Lake Ogallala; he's allowed us access to various ponds and wetlands for several years and talked about his water-witching experiences while sitting at a picnic table outside his store. The property along the river at Roscoe is owned by Darrel Thalken. In appreciation for Mr. Thalken's help we named a parasite after him; *Salsuginus thalkeni* is a worm that lives on the gills of the plains killifish.

My own doctoral adviser, Dr. J. T. Self, kindly wrote about his memories of Aute Richards and our visit to Tucson in a lengthy letter; Bob Kuntz added his comments in a telephone interview and also sent biographical materials. Dr. Self never loses his interest in former students and carries on a lively correspondence with many of us. Dan Brooks is, as of this writing, a faculty member at the University of Toronto and a scientific figure of international prominence. Mary Lou Pritchard is semiretired from her position as Curator of Parasitology at the University of Nebraska State Museum, and head of the Manter Lab. Mary Lou is also the archivist for the American Society of Parasitologists. Material on H. B. Ward came from the American Society of Parasitologists files. Skip is Mauritz C. Sterner who is collections manager in the Harold W. Manter Laboratory of Parasitology, University of Nebraska State Museum.

Charles Brown is a behavioral ecologist who received his Ph.D. from Princeton and worked for a number of years at Yale before taking a position at the University of Tulsa. He and his student helpers supplied the interview material at Cedar Point in the summer of 1993. Charles is a good friend, always willing to provide information and comments about a variety of biological topics—especially swallows—but he's also a very busy person in the summer. He was gracious enough, and his assistants courageous enough, to take time from their packed schedules to do the interview in Chapter 7. Charles also supplied the chapter summaries from his and Mary's book manuscript entitled, *Coloniality in the Cliff Swallow: Influence of Group Size on Social*

Behavior (Oxford University Press, due late 1994), which provided easy access to information he'd conveyed in seminars and published in a variety of scientific journals.

The woman who was in my first class is Karen Vierk; it was her son Byron's lizard that was dying while several couples were at a picnic at the home of a friend who lives in the country (introductory paragraph for Part II). Teresita Aguilar supplied the sheet music to "Malequeña Salerosa." The introductory paragraphs for Part III and Part IV relate instances with our son, John III and my wife Karen, respectfully; as of this writing John III is a graduate student in the Philosophy Department at the University of Nebraska–Lincoln and Karen is the Education Coordinator at the Sheldon Memorial Art Gallery as well as an instructor in the Museum Studies program. She spends more time studying, preparing for class, grading papers, and doing research than she spends making cookies.

The Road to Roscoe was originally written for a project by the photographer John Spence. During the middle 1980s, John did a remarkable series of Nebraska landscapes and asked me to write some text to accompany them. "Whatever comes to mind," he said. The only thing that came to mind was the South Platte at Roscoe and a feeling that I needed to make a definitive statement about that river. The photography and essay project never came to fruition; John did his work, I tried to do mine, no one was much interested in the results at the time. But I knew *The Road to Roscoe* would eventually see the light of day; no book of essays about the western High Plains wetlands would be complete without something on the braided rivers.

As for the literary sources: John Moore's book is *Science as a Way of Knowing: The Foundations of Modern Biology* (Harvard University Press, 1993). Some information in chapter 3 comes from Rachel Carson's *The Sea Around Us* (Oxford University Press, 1951), Estes et al.'s *Grasses and Grasslands: Systematics and Ecology* (University of Oklohoma Press, 1982), and John

McPhee's *Rising from the Plains* (Farrar, Straus, Giroux, 1986). The historical and dragonfly behavioral information in chapter 4 is largely from the doctoral dissertation work of Wendell Krull in the late 1920s. Krull's published work can be found in the *Transactions of the American Microscopical Society*, 50:215–277. Additional references used in chapter 4 include Arthur Koestler's *The Case of the Midwife Toad* (Random House, 1971) and Ellen Moore's *The Fairs of Medieval England: An introductory study* (Pontifical Institute of Mediaeval Studies, Toronto, 1985). Manter's Rules (chapter 6), along with an excellent discussion of the history of our ideas about parasite evolution, are from Brooks and McLennan's *Parascript: Parasites and the Language of Evolution* (Smithsonian Institution Press, Washington, D.C., 1993). The information on toxic quinones in tenebrionids (chapter 11) comes from Sokoloff's 1977 book *The Biology of Tribolium, with Special Emphasis on Genetic Aspects* (Clarendon Press), especially the chapter on *Tribolium* as a hazard and a food source! McColloch's papers on *Eleodes* appeared in the *Journal of Economic Entomology*, 11:212–224 and 12:183–194.

Ms. Cara Stanko helped a great deal with library research, general office work, and manuscript reading, for which I owe her many thanks.

Epilogue

Dunwoody Pond was written during the early 1990s, when my laboratory was filled with active young people searching for answers to big biological questions. Although my lab is still filled with such young people nearly a decade later, the faces have changed, and so have the questions. But two elements of this endeavor remain unaltered: namely, the focus on parasites, their restrictions to certain hosts, and the peculiarities of their life cycles and our dependence on Cedar Point Biological Station, with its surrounding rivers, streams, rolling sand hills, and prairie wetlands as a source of research problems.

As we did a decade ago, and indeed nearly three decades ago, when *Keith County Journal* was written, we constantly search for systems that can be used to address certain conceptual issues in parasitology. These systems are host-parasite combinations, which live in particular environments and have certain properties that allow the exploration of ideas about evolution and maintenance of a parasitic life. The conceptual issues are actually assertions about the general phenomenon of parasitism. This search for systems always requires that we get wet, dirty, and tired, that we spend countless hours at the microscope, and that we talk about our work not only in specific terms—counts, measurements, statistical analysis—but also, quite often, in artistic, literary, or metaphorical terms. Fortunately for a university professor for whom retirement is looming, the new faces seem willing to do both. This last fact strengthens my belief that our nation's future resides not so much in our young people alone but also in the way we interact with those young people—the way we seek to empower them instead of simply certify them, and the way we help them move intellectually from counts and measurements into the artistic, literary, and metaphorical implications of their work.

Virtually all of the people mentioned in the first edition of this book are now professional scientists in some capacity. Following the order of their recognition in "Notes and Acknowledgments," I know that: Aris Efting obtained her masters degree, chose not to pursue a Ph.D., and is now living in Savannah, Georgia; she does some consulting on water quality problems. Tami Percival is now a college professor; she received her Ph.D. in entomology from Texas A&M and married Jerry Cook, another entomologist; they both have faculty positions at Sam Houston State University in Huntsville, Texas. Laura Krebs is a doctoral student at the University of Arizona; she does not communicate often. Scott Snyder finished his Ph.D. at the University of Nebraska, won the American Society of Parasitologists' coveted Clark P. Read Young Investigator Award based on his UNL dissertation research, obtained a National Science Foundation postdoctoral fellowship to the University of New Mexico, and is now a biology professor at the University of Wisconsin-Oshkosh. Scott travels the world searching out parasitic worms. Richard Clopton finished a highly successful postdoc at Texas A&M and returned to Nebraska, where he is now a professor at Peru State College. Rich still publishes regularly, even from the small college laboratory setting; mentors student researchers; and sends them off to graduate programs elsewhere. Rich also took over the Natural History of Invertebrates course at Cedar Point Biological Station, which was taught for many years by Brent Nickol, another UNL parasitologist, winner of both the Distinguished Teaching Award and the American Society of Parasitologists' 2001 Mentorship Award. Obviously, Brent's shoes were relatively large ones to fill, but Rich did it quite successfully.

Of those who read the first manuscript of *Dunwoody Pond* for me, J. T. Self and Jane Geske are now deceased. Jill Anderson received her M.D.; she is taking a year's leave of absence to study human genetics at the National Institutes of Health. Mike Barger is a doctoral student at Wake Forest and is nearing completion of his Ph.D. Dan Holiday recently received his masters degree from Quinnipiac College in Connecticut. Dan hopes to return to Latin America and continue his study of parasitology and anthropology. Cara Stanko, a student helper in my office, is now in medical school; in her reading of the manuscript, she found one of the most truly embarrassing mistakes, one that would have branded me a complete literary incompetent, and for that discovery I am forever grateful. I wish she'd been around when some of my other books were written; they all have at least one truly embarrassing mistake, but only once has a person confronted me with one, and for that I am, as all writers in similar situations must be, quite grateful!

Speaking of errors, the original edition of *Dunwoody Pond* has pages 276 and 277 reversed, which was not my fault. That production mistake occurred in the middle of the last chapter, one of my more rebellious pieces of writing. When I went through my first copy of the published book, it was truly a frustrating experience to discover that page reversal. In rereading that chapter today, the page reversal is still frustrating, but it is somewhat overshadowed by the content. The narrative seems very close to the political-correctness boundary, although at the time it was written, everything in it felt like a fairly natural—even a rather logical—way of talking about not only our work, but the thoughts that were in our heads at the time we did that work. Granted, that chapter is a little introspective, in the sense of my explaining what it was that I was trying to accomplish with particular literary devices in *Keith County Journal*. Nevertheless, during the few short years since the book was published, the national cultural climate seems to have evolved enough to make me squirm a little bit and wonder who might be offended by what I'd written in 1992. Of course this author always hopes that someone might actually read what he wrote in 1992!

Of others mentioned in the original edition, whether in the text or acknowledgments, Mary Ann McDowell married Mike Ferdig; both received Ph.D.s from the University of Wisconsin-Madison, and both are now on postdoctoral research fellowships at The National Institutes of Health. Mike and Mary Ann easily made the transition from field work to molecular biology and immunology;

they now have two children; their first dog, Tenure, is deceased, but I hear they have another. Ralene Mitschler is now a professor at Western Maryland College; her husband Randy teaches at Hood College in Frederick. Tim Ruhnke is also a college professor, at West Virginia State University. Bill Moser, one of a long list of people who have been enthralled by leeches in western Nebraska, finished his masters degree at the University of Toronto and is now a scientist in the Division of Worms at the Smithsonian Institution. Bill probably represents the extreme case of a young person literally finding his life's work, and his intellectual sustenance, in a western Nebraska pond, and then parlaying that discovery into the one truly ideal job available anywhere in the world.

Lee Hardin is still a physician in the U.S. Army; my information indicates he's returned to Germany. Mary Lou Pritchard retired from her position as curator of parasitology at the University of Nebraska State Museum (UNSM) Manter Laboratory but still comes to work almost every day and is still the archivist for the American Society of Parasitologists (ASP). The ASP Web page is now maintained by the museum; Scott Gardner, a very successful nematologist, replaced Mary Lou as curator. Skip Sterner is still the collections manager at the UNSM Manter Lab. Bruce Lang is still a faculty member at Eastern Washington University. I could not locate Chris Rhodes, the minority student from chapter 5 whose test was misgraded; I sincerely hope he is doing well.

The local ranchers in Keith County, Nebraska, remain incredibly generous people. We have no way of thanking them adequately for the truly magnificent experiences we've had on their land except to say thanks, stop and talk periodically about what we found in their ponds, write books, and impress upon our students the extent of that local generosity. Well, there is one way we can thank such people after all: name new species after them, although nonbiologists, and indeed even some so-called modern biologists (or is it modern so-called biologists?), tend to laugh at the idea. Yes, undescribed species occur in Nebraska as well as almost everywhere else in the world. The first of our new species, *Salsuginus thalkeni*, a worm from the killifish's gills, was named for Darrell Thalken, who let us use his property along the South Platte River. The second of our new species, a gregarine protozoan, *Actinocephalus carrilynnae*, discovered in a damselfly, was not named after a rancher but after a little sister; its discovery and subsequent naming are described in chapter 2, "Choosing Damsels."

Sarah Richardson has finished her Ph.D. at the University of Arizona and, as of this writing, is looking for employment. Tami Percival's *Steganorhynchus dunwoodyi*, another gregarine from damselflies, is, of course, named for Duane Dunwoody. Recently, Megan Wise, a masters student, described yet another gregarine, this one from leeches, and named it *Metamera sillasenorum*, after the Jim and Lee Sillasen families, owners of the Nevens Ranch north of Paxton. During the summer of 2000, I took the Sillasens copies of the scientific paper bearing their name. Mrs. Sillasen laughed and said she'd read it; I assured her there would be no exam over the material.

Not long afterward, we met the new owner of another property we had used for years, which until recently had been part of an estate managed by a local bank. We were out collecting beetles along the road in the evening, and he came driving by.

"I'm Randy Peterson," he introduced himself from the cab of his pickup. It turned out he was a former student in one of my classes in Lincoln. "Keep doing what you're doing," he said. We had a long talk about biology and about college students. If you read the parasitological taxonomic literature in the next decade, watch for a new species with the specific epithet *petersoni*. I have no idea what the genus will be because we haven't found the parasite yet. But Randy is now on the list to be eventually honored, along with the McGinleys and Haythorns.

Scientific names, especially the specific epithets, last longer than many governments and certainly longer than professors' careers, and the journals that publish taxonomic work have much longer "shelf lives" than those specializing in the latest hot technology. "What is it?" will remain the most basic question in biology well into the future, perhaps even for centuries. Long after everyone in this book is gone, someone will wonder why all those western Nebraska parasites have honorific names. A little bit of checking will reveal the reasons, assuming there are still good libraries with scientific journals printed on real paper. That someone will then sit back and reflect on the ultimate value of high plains wetlands: they are classrooms of enormous power, and the people who go to school there are grateful almost beyond expression.

I hope, dear reader, that the above discussion of scientific names helps direct your eye toward labels on museum specimens and plants in public gardens. When the specific epithet—the second of the two words in a scientific name—seems to be someone's name but ends in -*i* or -*ii*, then you know, or can at least strongly suspect, that the epithet is an honorific. Behind that name is a story, most often a story of teaching, respect, and appreciation, an intellectual story, a story about shared good times, about exploration of natural history not only for its own sake, of course, but also for the sake of finding questions, the kinds of questions that occupy young minds like those described on the preceding pages, giving them their starts as professional scientists.

I've never known anyone to name a new species after someone he or she felt did not deserve the honor, the highest honor that a person with only deep knowledge about some often obscure part of the world can pay. Admittedly that feeling is certainly there at the time the naming decision is made; I don't know enough insider history to say for sure whether such feelings change over time, but I suspect, just from living in the midst of academic politics, that periodically describers of new species wish they'd not been so quick to lay on their honorific epithets. The Nebraska Sandhills ranchers who've opened up their property to us will, however, remain forever deserving. The link between their ponds and their nation's practicing scientists is clear and irrevocable; may there be many, many new species named after them.

The search for other Dunwoody Ponds has not been particularly successful. I'm convinced there are some, but I and a number of my graduate students have searched for them over and over again, with little real success. The Nevens Ranch overflow tank is as close as we've come to finding an equivalent laboratory, although Cedar Creek, east of Nevens, is proving itself to be extraordinarily complex, rich, and generous. Logic requires that if three such sites can be found along 15 miles of gravel road, then there must be thousands of them scattered across the high plains, and one needs only to keep looking. But the looking requires much time, energy, and permission for access.

I know such places exist in other parts of the country too because one of my colleagues at Wake Forest University, Dr. Gerald Esch, has found one. The place is called Charlie's Pond and is located in Stokes County, North Carolina. There have been several graduate theses produced from Charlie's Pond. Out of the seven papers published on work from there, none gives enough information so that a person could actually find the place. In all fairness, the papers concern parasite life-cycle processes and ecological relationships that likely characterize systems far from the shores of Charlie's Pond, but I strongly suspect that one day a student will ask Dr. Esch the age-old biology question—"What is it?"—and neither Esch nor the student will be able to come up with a satisfactory answer. At that point, the student may have to lay aside whatever plans he or she has for an ecological study and describe a new species. Then we'll find out where Charlie's Pond is, down to the nearest degrees, minutes, and seconds of latitude and longitude, probably read by some Global Positioning System (GPS) device. Then anyone, in theory, will be able to find it, and the species described from there as well. That, in theory, is the way field biology is supposed to work. I've never seen Charlie's Pond, but perhaps some day I'll try to go there, just to stand on the bank with an insect net and pluck a damselfly from nearby vegetation. Of course all of us in this business hope that when that day comes, we don't end up with our GPS unit in hand, reading latitudes and longitudes that are correct, standing where the banks of Charlie's Pond should be but instead dodging traffic in a strip-mall parking lot.

As for other characters in *Dunwoody Pond*, Charles Brown is still an active scientist, annually bringing a crew of young researchers to Cedar Point Biological Station. His second book, *Swallow Summer*, based on his western Nebraska experiences, was published by the University of Nebraska Press. I recommend it for another look into the minds of people who study nature and devise ways to get their questions answered but who must also interact with the general public and other scientists in order to accomplish their goals.

Dunwoody Pond ends with the story of a sandpiper flying in circles around me and some of my students, one of those truly unusual experiences that even a hardened scientist cannot explain away entirely by numbers and chemistry. Had I not written *Yellowlegs*, that day out at the Ackley Valley South marsh probably would not have registered very strongly on my emotional and—

yes, I almost hate to admit it—on my sort of spiritual, although natural spiritual, self. And had I not written *On Becoming a Biologist*, then another seemingly chance occurrence would not have had such an impact on me. I had gone to Dunwoody Pond in late May for the sole purpose of collecting two fish, one a bluegill and the other a largemouth bass. Neither fish had to be particularly large; a student in my lab, Megan Collins, had finished an excellent paper on the host distribution of gill parasites and because her paper was being submitted for publication, we had to save host specimens for deposition in the museum. The bluegill was not too difficult to obtain, of course, but the bass was being a little resistant to my lures. I heard footsteps behind me, through the vegetation, and turned to see a very excited young man with fishing gear hustling out in my direction.

"They said there was a marine biologist out here!" he said, with a big excited grin.

"Well, I am a biologist," I replied, "although not really a marine biologist."

Without waiting for much more of an explanation of who I was or what I was doing, he immediately launched into a discussion of the Dunwoody Pond biota, with special reference to the fish, rigging his pole as he talked almost nonstop. I must admit to being mightily impressed with the amount of knowledge an 11-year-old boy could acquire on his own, simply by studying a pond, his major research tool just a fishing pole.

Eventually we introduced ourselves. His name is Cody Wyckoff, and when he's not out in the field collecting something, he lives in Keystone, Nebraska. Through the summer of 2000, my students and I routinely encountered Cody, and one by one, the university students became increasingly impressed with his dedication to field biology and to science in general. Cody had an almost uncanny ability to see relationships between and the properties and fates of the organisms he pulled out of not only Dunwoody Pond, it turns out, but almost every other wet ditch he could find. My vision of scientists at the beginning of their career, the subject of this book, was expanded to include people a whole lot younger than university students.

I was not the only one suddenly amazed by Cody's research and teaching. Duane Dunwoody called one day.

"Doc, where can I buy a high quality loupe? I want to get one of those for Cody."

I looked up a company address and phone number and a catalog number for a small magnifying glass—the biologist's badge. After that conversation, I looked around the lab. A lifetime's worth of biological junk—strange chemicals, odd glassware, esoteric publications—decorated the shelves. Certainly there is something in here that would make a major difference in the life of an 11-year-old naturalist who lives next door to some high plains wetlands, I thought. From back in a storeroom there emerged a couple of old microscopes, ones that had not been used by anyone for years, and had I not rescued them for parts, they would likely have found their way into a landfill. I boxed them up and sent them out to Cody along with a letter. "You can return them when you come to the university," said the letter. If the athletic program can recruit, I reasoned, then so can the biology program.

This edition of *Dunwoody Pond* gives me the opportunity to tell this story. If writers have any dream, it is that their work will inspire others to behave in some particular way. My dream is that whoever has just finished reading this paragraph will sit back and think about all the young people he or she may have encountered, young people who are vitally and inescapably fascinated with nature, and do something to foster that fascination. When that happens, Dunwoody Pond will have done its job on people who have seen it only in their imaginations.